W9-CAJ-242

First published in the United States in 2003 by Grolier Educational, a division of Scholastic Library Publishing, Sherman Turnpike, Danbury, CT 06816

For Compendium Publishing
Contributors: Sandra Forty
Editor: Felicity Glen
Picture research: Peter Whitfield and Simon Forty
Design: Frank Ainscough/Compendium Design
Artwork: Mark Franklin/Flatt Art

Reproduced by: P.T. Repro Multi Warna, Indonesia.

Printed in China by: Printworks Int. Ltd

Library of Congress Cataloging-in-Publication Data
Whitfield, Peter, Dr.
History of science / Peter Whitfield
p. cm.
Includes index.
Contents: v. 1 Science in ancient civilizations – v. 2 Islamic and western medieval science – v. 3 Traditions of science outside Europe – v. 4 The European Renaissance – v. 5 The Scientific Revolution – v. 6 The eighteenth century – v. 7 Physical Science in the nineteenth century – v. 8 Biology and Geology in the nineteenth century – v. 9 Atoms and galaxies : modern physical science – v. 10 Twentieth-century life sciences.
ISBN 0-7172-5729-0 (act : alk. paper) – ISBN 0-7172-5703-7 (v. 1 : alk paper) – ISBN 0-7172-5704-5 (v. 2 : alk paper) – ISBN 0-7172-5705-3 (v. 3 : alk paper) – ISBN 0-7172-5706-1 (v. 4 : alk paper) – ISBN 0-7172-5707-X (v. 5 : alk paper) – ISBN 0-7172-5708-8 (v. 6 : alk paper) – ISBN 0-7172-5709-6 (v. 7 : alk paper) – ISBN 0-7172-5710-X (v. 8 : alk paper) – ISBN 0-7172-5711-8 (v. 9 : alk paper) – ISBN 0-7172-5712-6 (v. 10 : alk paper)
1. Science–History–Juvenile literature. [1. Science–History.] I. Grolier Educational (Firm) II. Title.
Q125 .W586 2003
509—dc21

2002029844

Acknowledgments
The publishers would like to thank the following for their help with the illustrations: Venita Paul and Sarah Sykes at the Science & Society Picture Library, Science Museum Exhibition Road, London SW7 2DD; and Arran Frood at the Science Photo Library Ltd. (**SPL**—www.science photo.com), 327–329 Harrow Road, London W9 3RB.

Picture credits
All maps and artwork are by Mark Franklin/Flatt Art.
All photographs were supplied by the Science & Society Picture Library except those on the following pages (T=Top; C=Center; B=Below): Author's collection p.44; SPL p.35 (David Parker), p.36 (Krafft/Explorer), p.43 (Tony & Daphne Hallas), p.46 (Space Telescope Science Institute/NASA), p.49 (Mark Garlick), p.52 (Mark Garlick), p.59 (Michael Dunning), p.70, p.71 (Mark Garlick), p.72 (B), p.73 (B) (NASA); p.8 (T and B—National Museum of Photography, Film and TV/SSPL); p.39, p.40, p.46, p.58, p.61, p.66–69 (all NASA/Science & Society Picture Library).

Note
Underlined words in the text of this volume and other volumes in the set are explained in the Glossary on page 70.

Contents

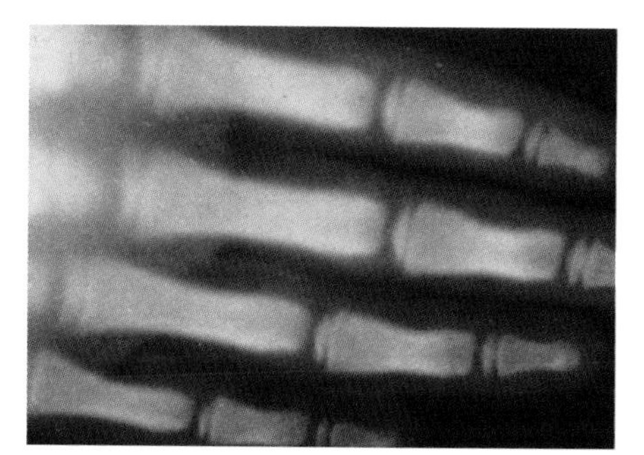

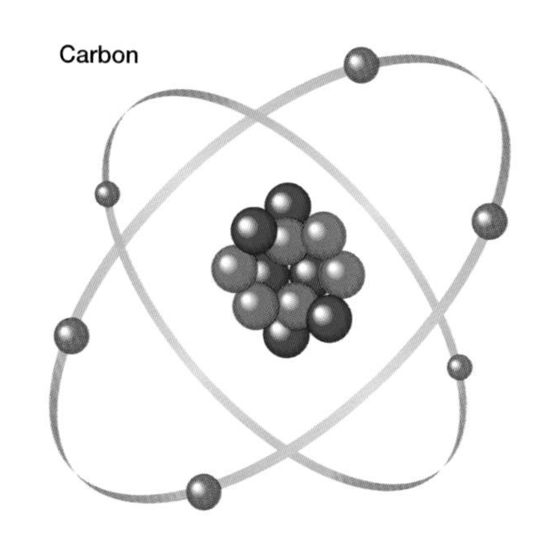

The Physical Sciences: Two Fundamental Problems

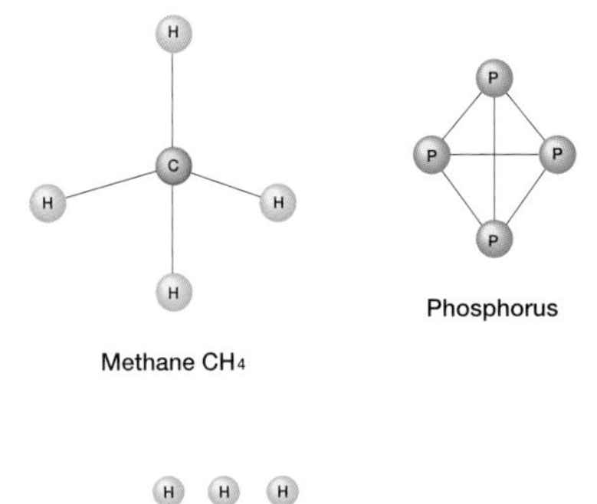

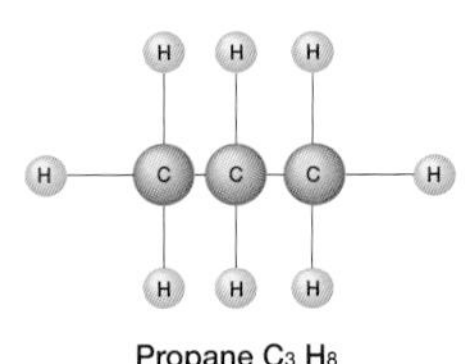

The atomic structure of methane, phosphorus, and propane.

The physical sciences of the twentieth century have addressed two fundamental problems: the nature of matter and the structure of the universe. In both cases startling new discoveries have brought about a revolution in human thought. The framework of ideas that scientists have hammered out in atomic physics and in cosmology seems securely based, and they are now universally accepted; but whether they represent the final truth it is impossible to say.

These two problems are of a very different kind. The structure of the universe is a purely intellectual challenge. How many billions of years ago did the universe begin? How many billions of miles does it stretch out across space? If the universe had a beginning, what existed before that? Does the universe go on forever, and if not, what lies outside it? These questions seem to have no practical bearing on human life, and yet we feel that they are critically important because they are the first step in answering how this Earth came to exist and hence why we are here.

Geologists study the formation of the Earth; biologists study the life forms on it; chemists study the composition of matter; physicists study the forces that organize matter. But all of these things have arisen in time; they had a beginning in which their natures were decided, and all their beginnings go back to one ultimate beginning. The origin of the universe is, in this sense, the fundamental scientific problem from which all others flow. Whether or not the problem can ever be solved, it is certain that it will continue to be posed.

NATURE OF MATTER

The problem of the nature of matter was also an intellectual challenge: It overturned established notions of space, time, and causality. Yet in addition, it has had huge practical consequences in human history. Nuclear power and electronic systems of communication and control—with all the consequences that they have had for social and political events—have flowed directly from the discoveries of the atomic physicists. The Industrial Revolution of the nineteenth century saw the emergence of science's role in shaping society, and it has been carried to dramatically new heights in the twentieth century.

Both physics and cosmology are baffling to the human mind because of the extremes of scale that they have uncovered: the infinitesimal scale of nature's basic particles and the inconceivable vastness of the universe. We wonder if it is an accident that our human world seems to stand midway between these extremes? Both science and logic seem to demand that these structures, from the smallest to the greatest, must be subject to immensely powerful organizing forces, and it is the attempt to understand these forces that forms the subject matter of physics.

SCIENCE AND POLITICS

The new cosmology took shape mainly in America and arose from discoveries made with powerful new telescopes, but important theoretical contributions came from Europe. The new atomic physics was very largely the creation of German scientists, and it arose at the time that so many other aspects of European culture were being uprooted and reshaped. Art, science, politics, and social norms were all revolutionized in the years around World War I (1914–18), and science too became implicated in this intellectual ferment. It is a striking fact that so many of the leading physicists were driven from Germany in the Nazi era, taking with them the skills necessary to build the first atomic bomb. Never before had pure science become so directly involved in political events. This episode was prophetic of the role that science would increasingly come to play in our society as it gradually invaded every aspect of our lives, sometimes for good and sometimes for ill.

Albert Einstein—one of the greatest original thinkers of all time.

Orion nebula (M42) from a photograph taken on February 26, 1883.

The Discovery of Radioactivity: The Curies

Above: Wilhelm Röntgen, discoverer of X-rays.

Opposite: Marie and Pierre Curie, pioneer researchers into radioactivity.

Below: Antoine Becquerel in his laboratory, around 1890.

The existence of atoms was a working hypothesis accepted by all physical scientists in the later nineteenth century. They were conceived to be the minute particles of which all matter was composed, joining with each other in accordance with complex rules of chemical bonding. But nothing at all was known about the atom itself, and there was no reason to suppose that it was anything other than an inert, unchanging building block. Between 1880 and 1920 this view of the atom was completely overturned, and one of the first indications of this approaching change came with the discovery of radioactivity.

One of the most important pieces of apparatus used by physicists at this time was the glass tube from which all or most of the air had been pumped out. It was used to study the spectrums of rarefied gases (see Volume 7, pages 52–55) and the way these gases behaved when an electrical current was passed through the tube. It was the behavior associated with these vacuum tubes that led to several far-reaching discoveries.

STRANGE ENERGY

In Wurzburg in 1885 Wilhelm Röntgen (1845–1923) realized that the tube with which he was carrying out some experiments was emitting some strange energy rays. These rays could not be particles, for they were not deflected by an electrical or magnetic fields, and they did not behave like normal light, for they were not reflected or refracted by lenses. They were related to light, for they exposed light-sensitive chemicals and made a kind of photograph. But the most astonishing thing was that they seemed able to penetrate solid matter: Röntgen was able to use them to take photographs of objects inside closed boxes and to show the human skeleton through the flesh, causing a scientific and public sensation. Unable to explain these rays, Röntgen called them X-rays, and physicists in laboratories throughout Europe began to study them. In Paris Henri Becquerel (1852–1908) found that similar rays were emitted by the element uranium, and this work was taken up by Pierre Curie (1859–1906) and his wife Marie (1867–1934), the most famous woman in the history of science.

The Curies knew that the richest natural source of uranium was the mineral pitchblend, but after prolonged analysis they also found that it contained two other elements they named radium and polonium. All these three elements were found to emit the form of

Imp

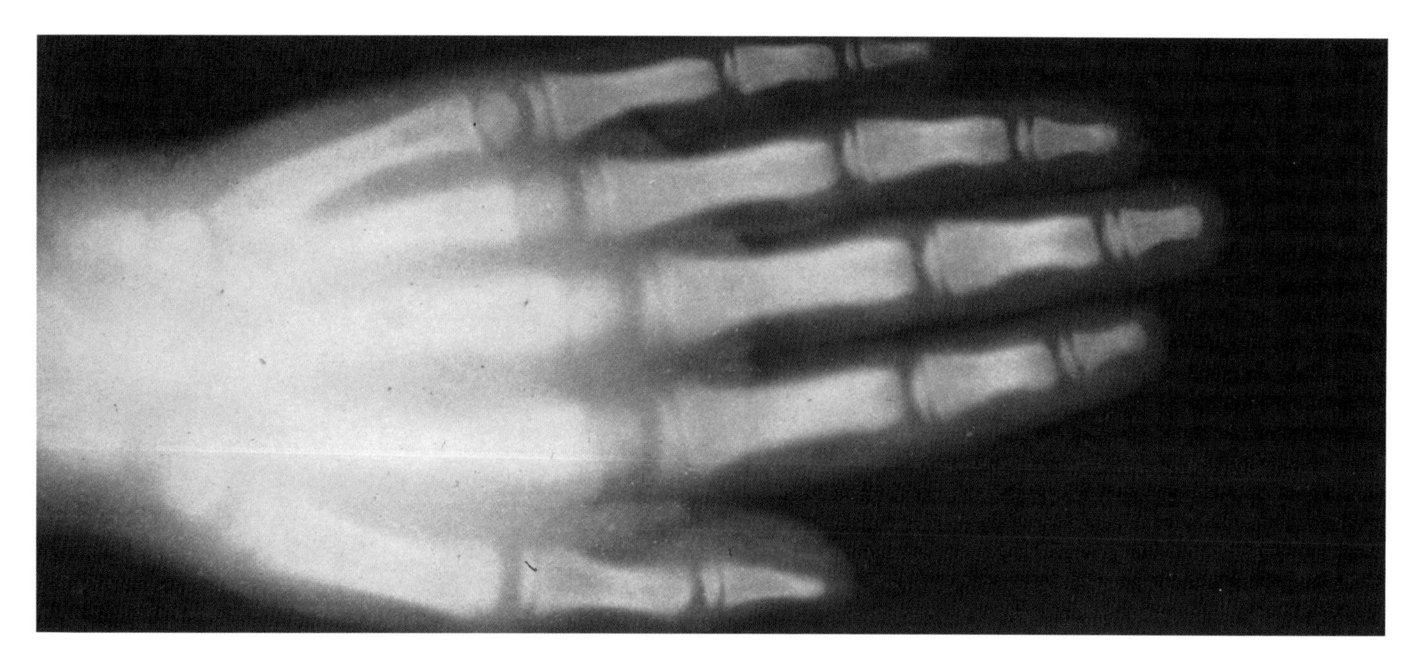

Above and Below: X-ray photographs made by Röntgen of the human hand and of a rat; these early X-rays amazed the scientific world and created a public sensation.

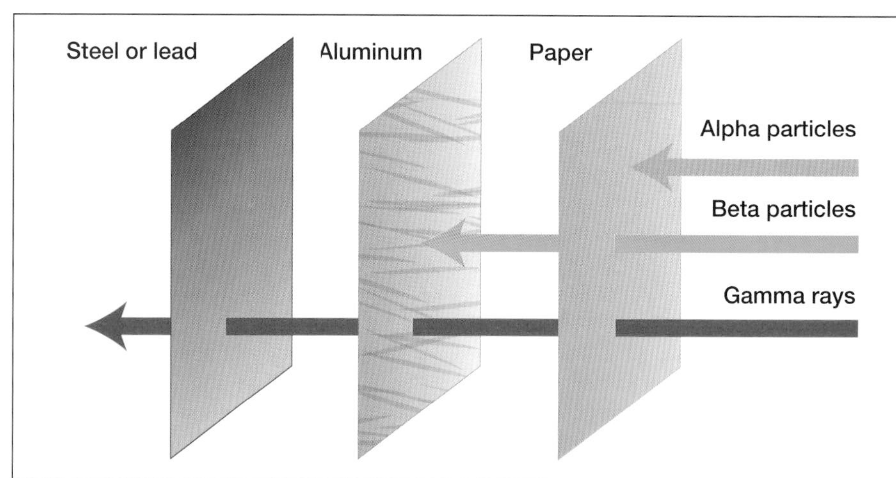

Right: After the Curies' initial discovery, radiation was analyzed into three main forms having different intensities.

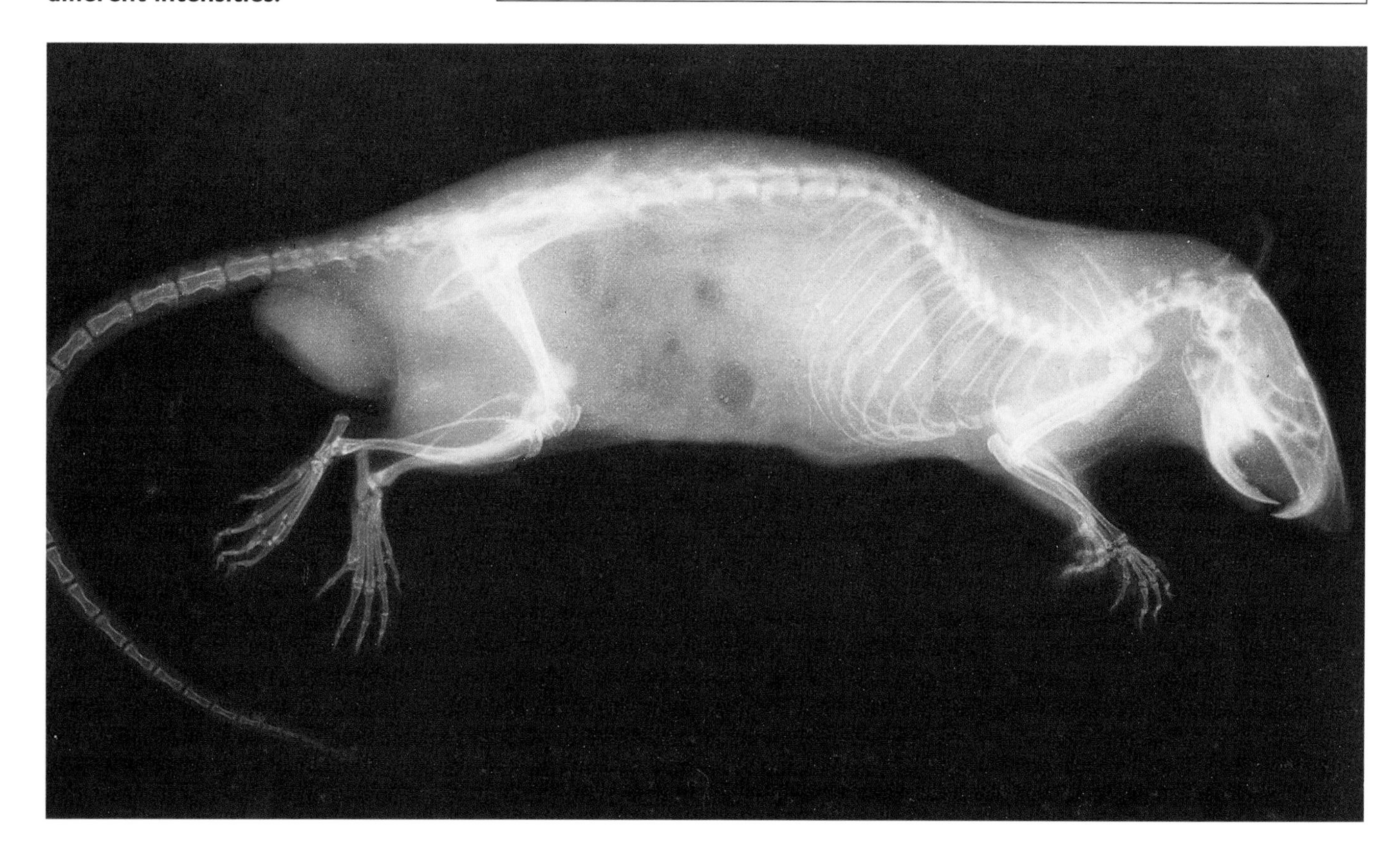

energy that they now termed radioactivity.

CHANGING ELEMENTS

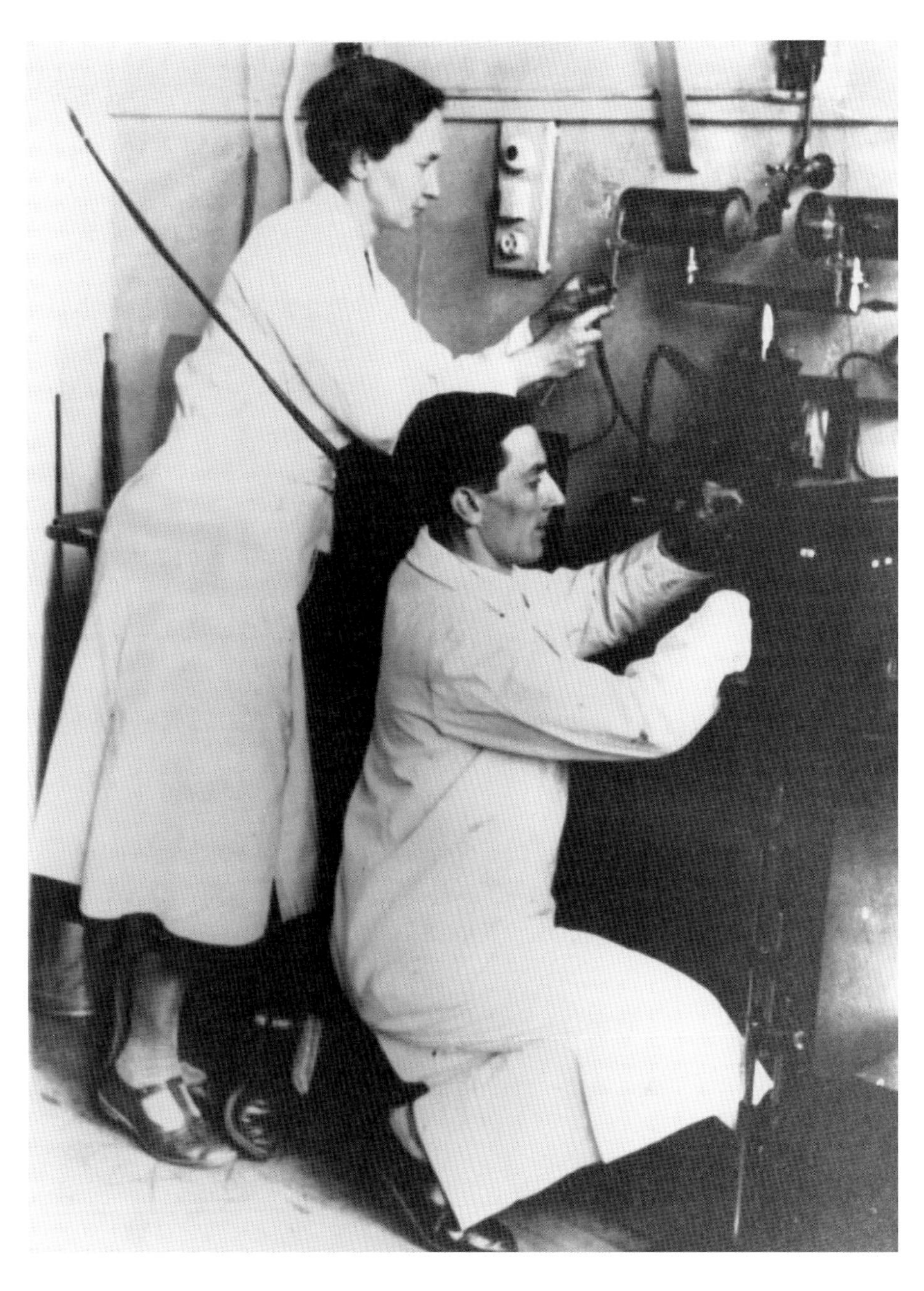

Above: Frederic Joliot and his wife, Irene Joliot-Curie, at work. Both would die of illnesses related to radioactivity.

Their next discovery was perhaps the most important. They tried combining radium and polonium with other materials and measuring the radioactivity of these various compounds, but they found that the level of radiation did not vary. The only thing that mattered was the amount of the element that was present. Therefore this special quality of radioactivity was not a type of chemical behavior but was characteristic of the element itself and of the atoms that composed it. Its ability to invade other elements must result from the emission of some form of force or energy. As this energy was emitted, the uranium itself appeared to decay and change its nature, eventually turning into common lead. But how was this possible? The elements, in classical science, were fixed and immutable; they were "solid, massy, and hard," in Newton's words, "even so very hard as never to break in pieces." And yet here were elements that were plainly unstable, spontaneously transmuting themselves one into another, and emitting rays from no known source that were able to penetrate solid matter.

The structural reasons for the behavior of the radioactive elements would be investigated by other physicists, but the Curies had overturned the accepted view of matter. Their research was carried out in quite primitive conditions, and they had no idea of the dangerous nature of the materials that they were investigating and would routinely handle lumps of raw uranium and radium. Both Pierre and Marie became ill, Marie dying of leukemia, although Pierre was killed in an accident unrelated to his work. The Curie's daughter Irene (1900–58) and her husband Frederic Joliot (1897–1956) both continued the research into radioactivity, and they too both died of its effects.

Atoms:
The Building Blocks of Matter

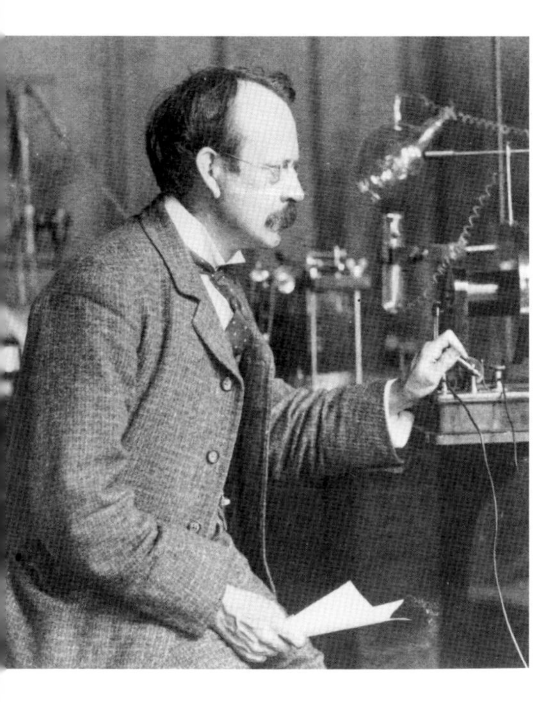

J.J. Thomson, discoverer of the electron.

JOSEPH THOMSON (1856–1940)

- Popularly known as J.J. Thomson.
- Pioneer of nuclear physics, discoverer of the electron.
- Born near Manchester, England.
- Educated at Owen's College, Manchester, then won a scholarship to Trinity College, Cambridge.
- 1884–1919 Held position of Cavendish Professor of Experimental Physics, during which time he turned it into the foremost research institute in the world.
- Studied electromagnetic theory and cathode rays.
- 1897 Published a paper in the *Philosophical Magazine* about cathode rays that revolutionized physics.
- Awarded the Nobel Prize for physics in 1906.
- 1908 Knighted for services to physics.
- 1911 Continued this research by studying positive rays, which led to his discovery of isotopes.
- 1914–18 Researched for the Admiralty and helped found the Department of Scientific and Industrial Research.
- 1918–40 Served as master of Trinity College.

The second great breakthrough made with the aid of the electrified gas tube was the discovery of the electron. In Cambridge between 1895 and 1897 Joseph Thomson (1856–1940) conducted a series of experiments from which he concluded that the rays that were emitted from the negative electrode in these tubes, the cathode, were not really rays at all, but were charged atomic particles. By placing a magnet near the tube, he established that these "rays" were deflected by the magnetic field. They must therefore be considered as physical bodies, and he found that they carried a negative electrical charge.

FUNDAMENTAL PARTICLE

Thomson knew the strength of the charge he was using and the degree of the deflection, and he was therefore able to estimate the mass of the particles that he was examining. He found to his astonishment that their mass was far less than one-thousandth of the mass of hydrogen, the lightest element known. Just as strange was the fact that the mass and charge of the particles were always the same no matter which gas was being used in the tube. Thomson was driven to the conclusion that some fundamental particle of matter had been discovered. In his own excited words this was "matter in a new state … this matter being the substance from which all the chemical elements are built up . . . . The carriers of electricity are bodies … having a mass very much smaller than that of the atom of any known element, and are of the same character, from whatever source the negative electricity derived." Thomson's word for this elementary particle was a "corpuscle," but this was soon replaced by the word "electron."

So the atom was not indivisible after all, but must be composed of even more minute particles, which under certain circumstances could be isolated. But why should matter at this elementary level be electrically charged, and why only negatively charged? What was the source of the counterbalancing positive charge? Was electricity truly diffused through space in the fields that Maxwell had described (see Volume 7, page 12), or was it concentrated into particles; and were charge and mass fundamentally aspects of the same thing?

MINUTE CALCULATIONS

When we talk about measuring and calculating the various aspects of the atom and the electron, this was of course achieved in purely mathematical terms. No one had yet seen an atom or been able to describe it as an individual entity; but scientists such as Maxwell,

Kelvin, and Boltzmann (see Volume 7, pages 8–12) had developed a mathematical language with which electrical charges or the kinetic movement of gases could be carefully analyzed and related to the atomic weights that chemists had already established. Already in 1865 the Austrian chemist Joseph Loschmidt (1821–95) had used gas kinetics (movements) to calculate the theoretical size of an atom: His figure of 10^{-7} (0.00000009mm) is remarkably accurate. In 1908 the French physicist Jean Perrin (1870–1942) sought a more concrete demonstration of the reality of atoms.

Perrin achieved this by analyzing what scientists called "Brownian motion," named after the Scots botanist Robert Brown, who had described it as early as 1828. What Brown noticed was the way in which tiny particles in water, such as pollen grains, were in continuous motion. Brown could not explain this, but it was later understood to be the result of the thermal motion of the water: Molecules of water were continuously bumping into these particles and moving them around. Perrin's argument for proving that atoms had true, measurable mass was based on an analogy between these particles in a liquid and the molecules in the atmosphere. The thinning of air at high altitude depends on the opposing forces of gravity, pulling the molecules down, and thermal motion forcing them up. The relationship between the weight of the molecules and the height of the atmosphere would be the same, Perrin believed, for the Brownian particles suspended in water. Perrin counted the number of particles of various substances found at different heights in water samples, and from this he calculated the number of atoms and their rate of decrease. The numbers were directly proportional to the atomic weight of the substances, and they confirmed Loschmidt's theoretical results.

After Perrin's experiments were announced in 1908, no scientist continued to doubt the physical reality of atoms. But now it was clear why even the most powerful microscopes could never hope to reveal individual atoms: They were far smaller than the wavelength of light itself. So, in just a dozen years between 1895 and 1908 the scientific community had seen the cherished doctrines of the immutability and indivisibility of atoms overturned. But it had also found new techniques for probing the physical reality of these, the minutest building blocks of all matter. The next few years would see even more astonishing ideas put forward concerning the structure and properties of atoms.

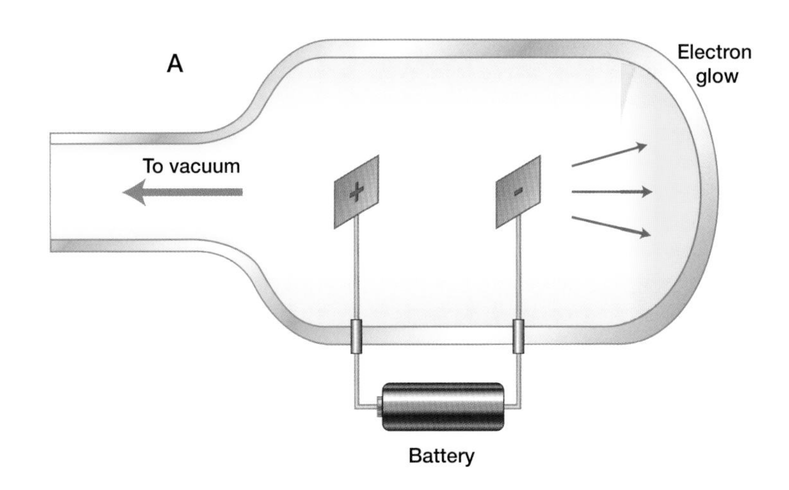

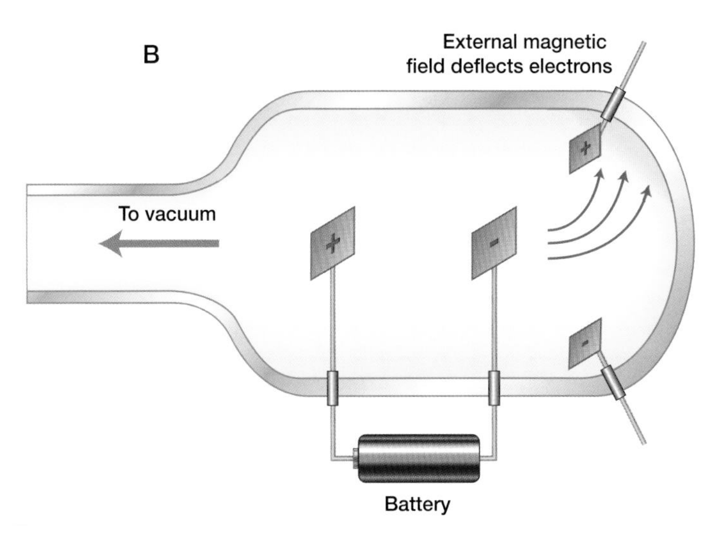

Thomson's cathode-ray tube experiment. He was certain that these rays were physical particles because they could be deflected by a magnetic field.

Robert Brown (1773–1858), Scottish plant taxonomist who gave his name to Brownian motion.

The Structure of the Atom: Ernest Rutherford

Ernest Rutherford (right), the father of atomic research.

To rationalize the strange new facts about the nature of matter that Thomson and the Curies had uncovered was the achievement of Ernest Rutherford, a New Zealand-born physicist who carried out his work first in Canada and then in Britain. Rutherford (1871–1937) turned his attention to radioactivity when he studied with Thomson at the Cavendish Laboratory in Cambridge in the 1890s, and he went on to analyze both the radiation, which he showed to consist of several different components, and what was happening to the radioactive substance itself.

The second point was of the deepest interest, for Rutherford found that the radioactive element changed its properties as it emitted radiation and turned into a related but distinct substance with slightly different properties. He found that radium decayed into the gas radon, and the element thorium decayed over a fixed period of time into a series of other elements, finally stabilizing into a form of lead. Since chemical theory defined each element as unique in its atomic nature, it was clear that radioactivity was an indication of some fundamental change inside the atom. So the atom was not immutable, and it must have its own complex structure.

Opposite: The solar-system model of the atom proposed by Rutherford, the electrons orbiting the nucleus as the planets orbit the Sun.

MINIATURE SOLAR SYSTEM

Rutherford's most historic experiments were conducted at Manchester University in 1910. Here he directed a beam of radioactive particles from the gas radon at extremely thin gold foils, expecting them to penetrate the foil like X-rays. Rutherford described what happened as "quite the most incredible event that has ever happened to me in my life." Most of the rays did pass through, but a very small proportion were deflected and bounced back. He described his reaction in words that became famous in scientific history: "It was almost as incredible as if you fired a 15-inch shell at a piece of tissue paper, and it came back and hit you."

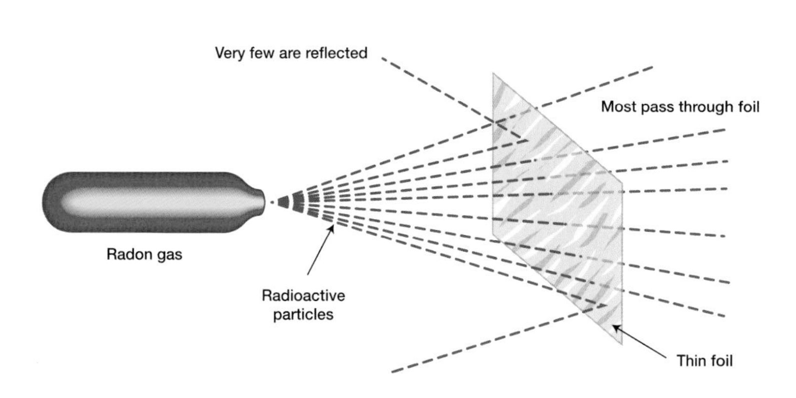

Rutherford's famous foil experiment, which led to his theory of atomic structure.

Pondering on this strange result, he came to the brilliant conclusion that most

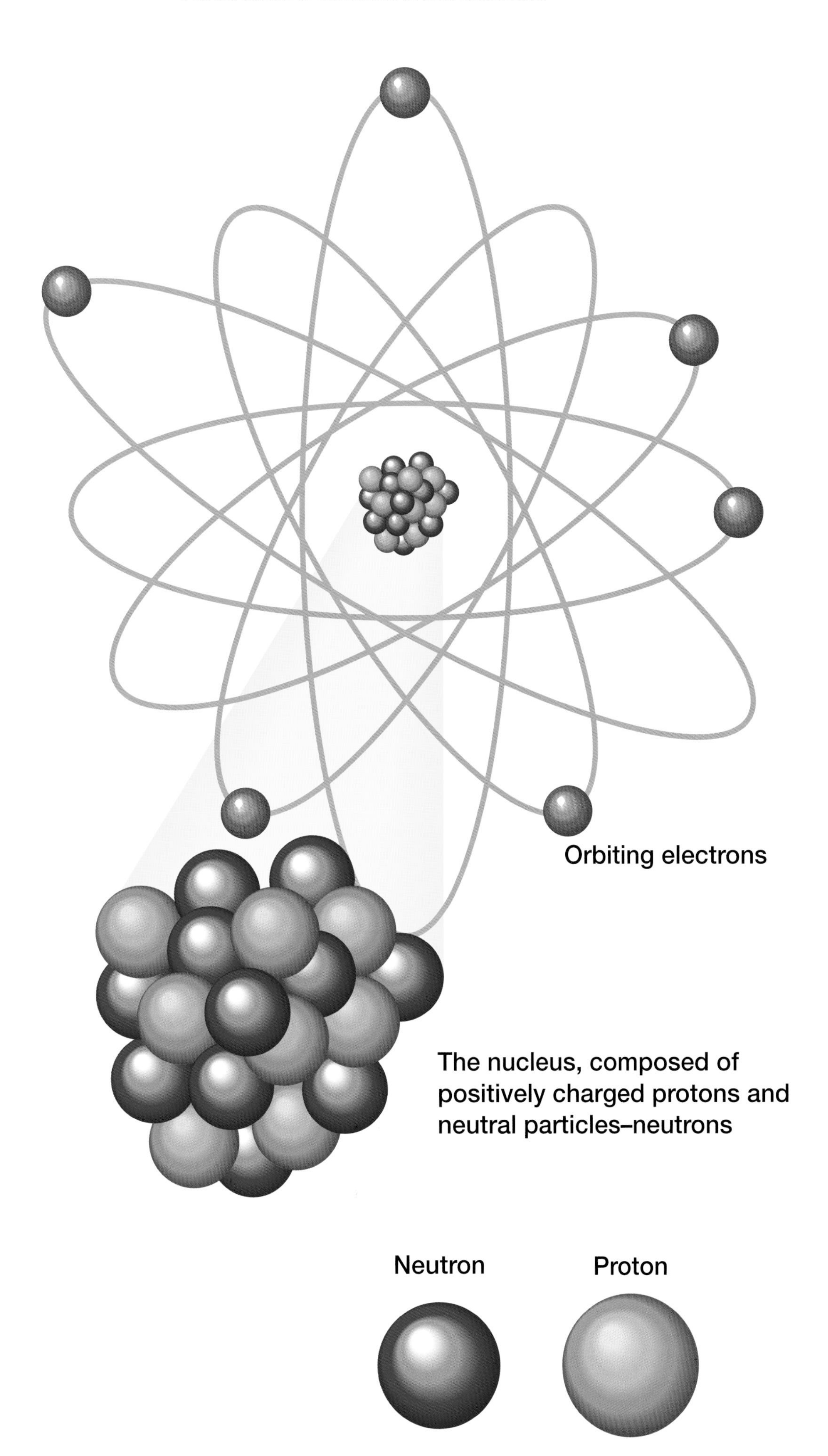
Orbiting electrons
The nucleus, composed of positively charged protons and neutral particles–neutrons
Neutron
Proton

Above: Portrait of Rutherford.

Opposite: Rutherford in the Cavendish Laboratory in Cambridge, the historic center of early atomic research.

ERNEST RUTHERFORD (1871–1937)

- Father of nuclear physics.
- Born Brightwater, New Zealand.
- Won scholarships to Nelson College and then Canterbury College, Christchurch.
- 1894–96 Researching into magnetism.
- 1895 Won a scholarship to Cavendish Laboratory, Trinity College, Cambridge.
- Made first successful wireless transmission over two miles.
- Discovered three types of uranium radiations.
- 1898 Appointed professor of physics at McGill University, Canada.
- With Frederick Soddy formulated the theory of atomic disintegration.
- 1907 Moved to Manchester University.
- Invented a radiation counter with Hans Geiger.
- 1908 Won the Nobel Prize for chemistry.
- With his assistant Niels Bohr developed the concept of the "Rutherford-Bohr atom."
- 1919 In a series of experiments he discovered how to liberate hydrogen nuclei.
- Moved to Cambridge to make it the world center for the Newer Alchemy in 1919.
- 1920 Predicted the existence of the neutron.

of the atom's mass was concentrated into one small region, while the rest was empty space. The tiny, light, negatively charged electrons that Thomson had discovered were at a distance from the nucleus, which was more massive and carried a positive charge. It was like a miniature solar system, but the scale that Rutherford calculated was truly astonishing: The size of the central nucleus, he suggested, must be around 10,000 times smaller than the diameter of the whole atom. The reflected particles in his experiment had struck this nucleus, while the rest had passed through the empty space.

Early accounts of this solar-system atom tried to convey the scale involved by picturing the nucleus as a pinhead in the midst of a vast cathedral, while the electrons were orbiting high in the vast dome. These outer particles were revolving so swiftly, at about one-tenth of the speed of light, that the atom as a whole behaves like a solid in the same way that the spaces between the blades of a high-speed turbine vanish for all practical purposes.

SCIENTIFIC SENSE

This model of the atom was later to be modified in some important ways, but it has remained a linchpin of modern science. Later research with atoms of different elements revealed that they have different numbers of electrons and particles in their nuclei. Suddenly the weights of the elements in the periodic table made scientific sense: The characteristics of the related elements derived from their atomic structure. To change the number of atomic particles would be to change one element into another, and this is exactly what Rutherford did when he disintegrated atoms of nitrogen to produce oxygen. The alchemist's dream of transmuting the elements became a reality in his Cambridge laboratory in 1919.

It was Rutherford who saw the answer to a puzzle that had troubled the scientific community for 50 years—that of the age of the Earth (see Volume 8, page 19). Physicists such as Kelvin had argued that the Earth could not possibly be as old as geologists such as Lyell had claimed because it must have cooled in a few million years after its creation. Rutherford was able to explain that radioactive elements within the Earth had provided sources of heat that had endured for billions of years—far longer than the cooling process understood by preatomic physics. Rutherford possessed a clear sense that in probing the structure of the atom, he was peering into the seeds of the universe. "When we have found how the nucleus of the atom is built up," he wrote, "we shall have found the greatest secret of all—except life. We shall have found the basis of everything—of the Earth we walk on, of the air we breathe, of the sunshine, of our physical body itself, of everything in the world, however great or however small—except life." Perhaps more than any other person, Rutherford deserves the title of father of the atomic age.

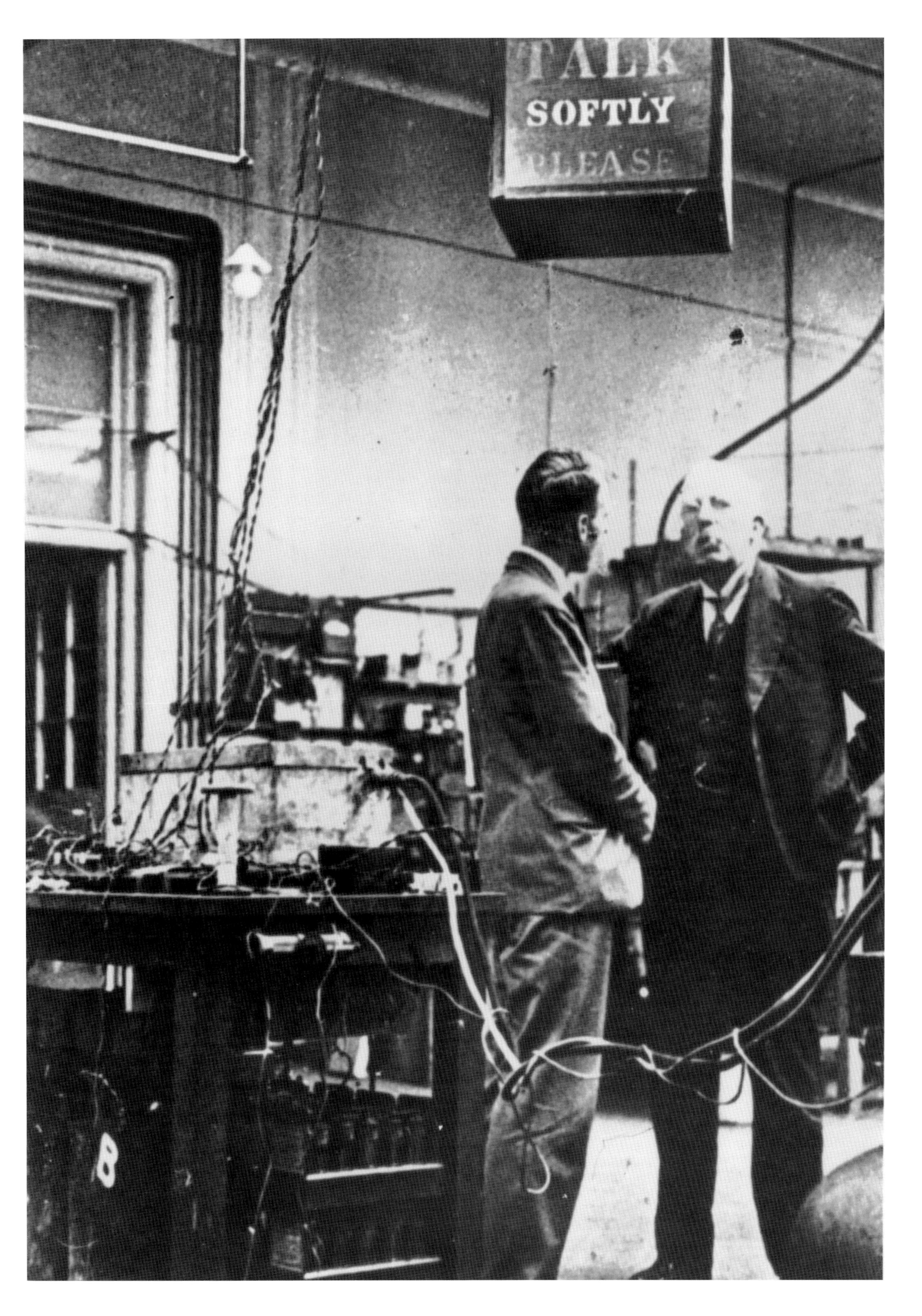
TALK
SOFTLY

The Quantum: Max Planck

The further development of Rutherford's model of the atom's structure was made possible by one of the most far-reaching advances in modern science—the discovery of the quantum. Quantum theory is now central to the physical sciences, but its workings and its implications are strange and puzzling, and more than three decades were needed for international scientists to agree about its validity.
The founder of quantum physics was Max Planck (1858–1947), a deeply cultured man from northern Germany who excelled at music and classical science and languages, but who saw scientific study as the only route to the absolute truth.

HOT BODIES

Planck's most important work was carried out in Berlin in the 1890s and centered on the problems of measuring the radiant energy—the heat and light—given off by a heated body. It is well known that heated or burning bodies emit a red glow that turns to orange, then to yellow, and finally to white, the wavelength of the light becoming shorter as the temperature increases.

For many years physicists had tried to formulate a law stating exactly how the amount of radiant energy given off in this way varied with wavelength and temperature. All attempts had failed because it was naturally assumed that this radiation could rise and fall in a steady curve and could have any value from zero to infinity. Instead, Planck found that he could arrive at a law relating energy to wavelength only by assuming that energy is emitted not in an unbroken stream, but in discontinuous bits or portions, which he termed a quantum, Latin for "how much."

MAX PLANCK (1858–1947)

- Theoretical physicist.
- Born Kiel, Germany.
- Studied at Munich, then Berlin universities.
- 1889–1926 Professor of physics at Berlin University.
- 1900 Formulated quantum theory after working on thermodynamics and black body radiation.
- 1918 Awarded the Nobel Prize for physics.
- 1930 Elected president of the Kaiser Wilhelm Institute (renamed Max Planck Institute after the war). Resigned in 1937 in protest against the Nazi regime.

PLANCK'S CONSTANT

On purely mathematical grounds he concluded that each quantum carried an amount of energy given by the formula $E = hv$, where v is the frequency of the radiation, and h is Planck's Constant, a tiny but invariable number that has since been proved to be one of the most fundamental constants in nature. In any process of radiation the amount of energy emitted divided by the frequency is always equal to h. The Planck's Constant is often referred to as the elementary quantum of action, the smallest measurable event in the physical world. Its value is 0.000000000000000000000000006624. Although Planck's Constant has dominated the computations of atomic physics for almost a century, it cannot be explained, any more than the speed of light can be explained: It is simply a basic and invariable fact of nature. Planck published his first paper on the quantum in December 1900, thus inaugurating a new era in the history of science.

Why is this tiny number so important? Planck himself was not at first involved in investigating the structure of the atom, and it was left to others to explore the implications of quantum theory—to make the crucial connection between the infinitesimal packets of energy of which radiation events were composed and the way the infinitesimal building blocks of matter behaved. The gradual revelation that quantum theory could act as a powerful model for analyzing mass, charge, and momentum within the atom showed that mass and energy were intimately connected and provided a new language for atomic physics.

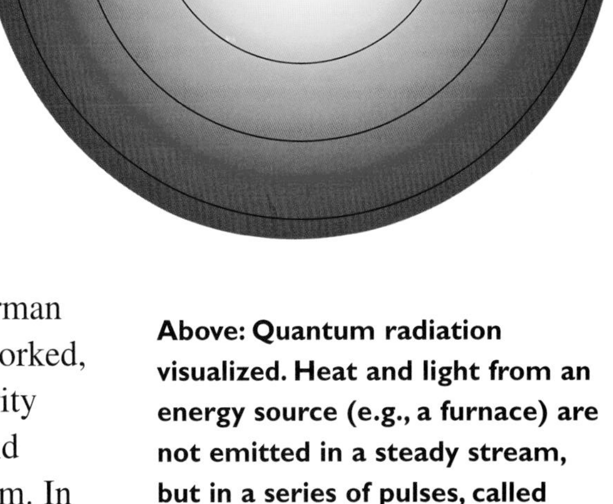

Above: Quantum radiation visualized. Heat and light from an energy source (e.g., a furnace) are not emitted in a steady stream, but in a series of pulses, called "quanta" represented here by the pulse circles.

Opposite: Max Planck, discoverer of the mysterious quantum effect, which revolutionized atomic physics.

This revolution made Planck into the senior figure of German science. The Kaiser Wilhelm Institute in Berlin, where he worked, was later renamed the Max Planck Institute. Planck's authority stemmed from his personal moral character—his honesty and integrity impressed everyone who came into contact with him. In the 1930s he went personally to Hitler to argue against his racial policies, and he decided to remain in Germany throughout the Nazi era to attempt to preserve what he could of German culture. Only his immense prestige kept him safe during these years, but it could not his save his son, Erwin, who was implicated in the July 1944 plot to assassinate Hitler and was then executed.

Quantum and the Atom Structure

Right: Niels Bohr, the Danish physicist who used the new quantum theory to build a more accurate but more complex picture of the atom.

Far Right: Bohr's picture of the electron moving into different orbits around the nucleus, emitting or absorbing energy as it does so.

NIELS BOHR (1885–1962)

- Physicist.
- Born Copenhagen, Denmark.
- Attended Copenhagen University, then moved to Cambridge to work with J.J. Thomson and Manchester to work with Rutherford.
- 1913 Greatly extended the theory and understanding of atomic structure by explaining the spectrum of hydrogen using an atomic model and quantum theory.
- 1916 Became professor of physics at Copenhagen University.
- 1920 Founder and first director of the Institute of Theoretical Physics, Copenhagen.
- Awarded the Nobel Prize for physics.
- Fled occupied Denmark in World War II for America, where he did atomic bomb research.
- 1945 Returned to Copenhagen to work on nuclear physics; developed the liquid drop model of the nucleus.

"When it comes to atoms, language can be used only as in poetry: The poet too is not nearly so concerned with describing facts as with creating images." These were the words of Niels Bohr, the Danish physicist who brought together Rutherford's vision of the atom as a miniature solar system and the quantum mathematics of Max Planck. Bohr (1885–1962) emphasized that the structure of the atom strained human powers of observation and measurement to the limit. Scientists such as Rutherford had suggested various ways in which atoms might be visualized, but their nature is so elusive that what atoms really are defies understanding, so that their behavior can only be described mathematically. Nevertheless, Bohr refined and reinforced the solar system picture of the atom so successfully that it is still the Rutherford-Bohr model that dominates our minds.

ELECTRON JUMP

The great objection to Rutherford's model of the atom was this: Unlike the planets revolving around the Sun, the electrons carried a

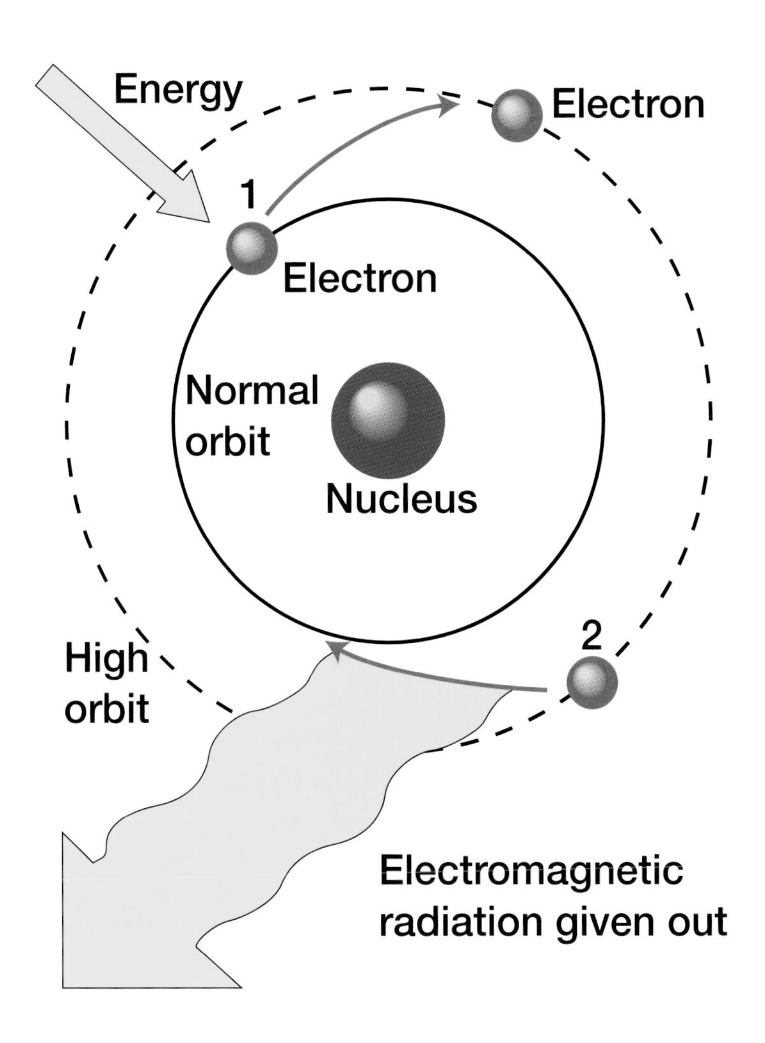

negative charge and were evidently attracted to the positively charged nucleus of the atom. In that case, why did not the electron spiral down into the nucleus? Why was the atom a stable entity? Bohr answered this problem in papers published in 1913 by using the newly discovered quantum principle: The electron, he argued, only circled the nucleus in certain predetermined orbits. When an atom was exposed to energy in the form of heat or light, the electron absorbed that quantum of energy and jumped to a higher orbit; afterward it fell back to a lower orbit, emitting the same quantum of energy. The infinitesimal quantum of energy that Max Planck had discovered was precisely the measure of energy involved in the electron jump. Electrons could not orbit the nucleus in any paths they chose, for this was the force of the quantum model: They could only process energy in discrete packets to jump from one predetermined orbit to another. The electron's innermost orbit was fixed, so that it could not spiral down into the nucleus: The classical solar system mechanics did not apply to the atom. So matter and energy were

Proton
Neutron
Electron
Hydrogen
Helium
Lithium
Carbon

The stability of the atom. The number of protons, neutrons, and electrons always balance each other.

Right: Chemistry becomes physics. Bohr's theories helped explain the chemical bond between elements in a compound as the sharing of electrons.

Opposite: James Chadwick completed the classical understanding of the atom with his discovery of the neutron; an article from *The Times* of London.

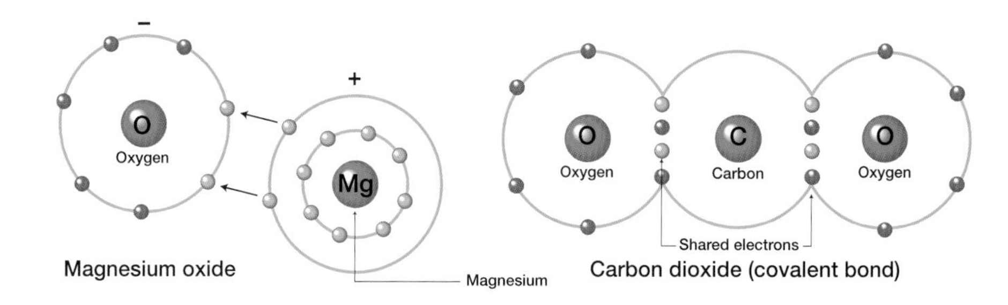

intimately related at the most fundamental level: Both the absorption and the emission of energy produced changes in the atomic structure of matter. The atom was a storehouse of energy, and its fluid structure made possible the conservation of energy, which was one the great principles of nature discovered by nineteenth-century science.

GHOSTLIKE WEB

The common-sense understanding of matter as solid, hard, and unyielding was replaced by a vision of matter as a ghostlike web composed largely of empty space in which energy can be received, stored, and emitted. The speed and scale on which these quantum events occur make it impossible to build a true visual model of the atom: Its behavior can only be described in aggregates (collectively), in mathematical patterns.

Below: Niels Bohr and family.

Bohr's work was the starting point of the mature quantum theory of atomic structure. Some of his successors proposed other very different models for understanding the relationship between the electron and the atomic nucleus, but Bohr did not disagree with them, for it was the mathematical patterns that were all-important. In one sense the core subject matter of chemistry had moved into the realm of physics, for the structure of nucleus and electrons now illuminated the long-standing problem of the chemical bond. Those electrons farthest from the nucleus were, in many elements, quite weakly attached to the nucleus and could be captured by an atom of another substance. So the bond between two separate elements that had come together to form a compound consisted of shared electrons.

SPECTRAL ANALYSIS

The idea of quantum energy inside the atom also explained some very strange results from the spectral analysis of the invisible rays of light and radiation given off by gases. In the 1880s the Swiss mathematician Johann Balmer noticed that the black lines in the hydrogen spectrum showed a curious pattern of spacing, with each line spaced apart by an exact multiple of a constant—1c, 2c, 3c, 4c, and so on. No one could explain this regular pattern; but if the atoms of hydrogen were radiating energy at discrete levels as Bohr claimed, then the spectral patterns could be seen as a precise image of that activity.

Other physicists extended this approach to other elements and found that it held true, each having its own characteristic pattern. Therefore the idea of quantum energy active within the atom was clearly established.

Further evidence of the individual structure of different elements was provided by the discovery of X-ray crystallography in 1912 by the German physicist Max von Laue (1879-1960). When a fine beam of X-rays is passed through a crystal of any substance, a photograph can be obtained indicating the regular arrangement of atoms in the crystal. It is not a direct photograph of the atoms, but a refraction of the X-rays themselves in a pattern that is specific to each substance. These developments shed new light on the periodic table of the elements, for it became clear that the relationships within that table were structural relationships at the atomic level: Similar atomic structures produced similar chemical properties.

This may seem self-evident now, but it was not so before Rutherford and Bohr. It was experimental work with particles that led the British physicist James Chadwick (1891–1974) to refine the picture of the atomic nucleus. The puzzle concerning the nucleus was that it appeared to be exactly twice as heavy as could be accounted for by protons alone. Chadwick showed that the nucleus consisted of a proton, which was positively charged, and a neutron, which had the same mass as the proton but no charge

A NEW RAY

DR. CHADWICK'S SEARCH FOR "NEUTRONS"

Dr. James Chadwick, F.R.S., Fellow of Gonville and Caius College, Cambridge, in an interview on Saturday said that the result of his experiments in search of the particles called "neutrons" had not, at the moment, led to anything definite, and the element of doubt about the discovery still existed.

There was, however, a distinct possibility that investigations were proceeding along the right lines. In that case a definite conclusion might be arrived at in a few days, and on the other hand it might be months. Dr. Chadwick described his experiments as the normal and logical conclusion of the investigations of Lord Rutherford 10 years ago. Positive results in the search for "neutrons" would add considerably to the existing knowledge on the subject of the construction of matter, and as such would be of the greatest interest to science, but, to humanity in general the ultimate success or otherwise of the experiments that were being carried out in this direction would make no difference.

Lord Rutherford, at the conclusion of his lecture at the Royal Institution on Saturday, confirmed in a statement to a representative of *The Times* the importance of the experiments by Dr. Chadwick. They seemed to point, he said, to the existence of a ray whose particles, known as "neutrons," were indifferent to the strongest electrical and magnetic forces. He did not, however, confirm the assumption that these particles were matter in the everyday sense of the word from the fact that their collisions obeyed the laws of momentum; nor the further assumption that they moved at a speed more than a tenth of that of light or electricity.

Lord Rutherford's own lecture dealt with the "Discovery and Properties of the Electron," and in the course of it he conducted a number of experiments in the generation of cathode rays, with glass tubes and bulbs used 50 or more years ago by Sir William Crookes, and with others lent by Sir William Pope.

The discovery in 1897, said Lord Rutherford, of the negative electron had been of profound significance to science. The electron tube was essential to-day not only for the generation of continuous radiations, but for their reception, and thus rendered possible the rapid development of radio-telephony and broadcasting. When they looked back, it became clear that the year 1895 marked what they might call the definite line between the old and the new physics. It was in that year that Röntgen made his famous discovery of the X-rays. The importance of the discovery was really, in a sense, greater than itself, because it gave the impetus to experiments which led to two epoch-marking discoveries, of radioactivity by Becquerel in 1896 and of the electron in 1897. Those two discoveries opened up new vistas of the wonderful ways in which Nature worked. They gave us, for the first time, methods for attacking the question of the structure of the atom, and we now had a fairly definite general view of that structure; and they had given us, for the first time, a fairly clear idea of the mechanism of radiation.

MARCH SESSION AT OLD BAILEY

TWO MURDER CHARGES

Waves, Particles, and Mysteries

Bohr's model of the atom was easy to visualize in the case of a light element with few electrons—hydrogen, for example, has only one. But it was not flexible enough to deal with the heavier atoms that have many electrons, especially with extreme cases such as uranium, with 92. In the 1920s quantum physics made dramatic progress by embracing the idea that radiation, including the energy of the electron, behaves simultaneously like waves and like particles. Light, for example, having been described as a wave in classic electromagnetic theory throughout the nineteenth century, was now found to behave as if it were discrete quanta—individual bits—of energy, termed photons, that produced measurable physical effects.

DUAL NATURE

The idea that the energy within the atom could have a dual nature—sometimes behaving like waves and sometimes like particles—was put forward by the French physicist Louis de Broglie (1892–1987) and was given a new mathematical form by the Austrian Erwin Schrödinger (1887–1961). Schrödinger's wave equations of 1926 treated electrons as waves that increased in energy and frequency as they moved further away from the nucleus. A complex, heavy atom such as uranium now appeared not so much as an orderly solar system as a cloud of electrons. It became doubtful whether the exact physical position of an electron could ever be defined. This problem was made explicit by the German physicist Werner Heisenberg (1901–76), who claimed that the individual atom could only be analyzed in terms of probability. Heisenberg insisted that the baffling fact of the quantum leap made it necessary to abandon completely the attempt to visualize the atom and to think of it only mathematically.

There is a very simple reason why this has to be so: No test can be devised to show the position of the individual electron, for any observation must entail some input of energy that will cause its position to move. If we could shine some beam of ultra short-wave light in order to illuminate the atom, the energy of that beam would have a dramatic effect on the electron. We can therefore never see the atom in its natural state. Like a coin spinning in the air, the question of heads or tails can only be determined by interfering radically with its movement. To the question why does physics employ such mysterious modes of description, the physicist answers, because they work.

Louis de Broglie.

ULTIMATE ELEMENTS

In the laboratory the behavior of atoms en masse allows us to infer what is happening at the level of the individual atom. The aim of the physicist is to state the laws of nature in ever more precise mathematical terms. The first historic proof of the success of this work was the detonation of the atomic bomb in 1945. So the abstract language of mathematics can describe how atoms behave without being able to tell us what they look like, still less what they really are. In the last 50 years research has revealed a complex world of subatomic particles that are pure charges of high energy, and whose function is to bind together atomic nuclei. These subatomic particles, or quarks, seem to be the ultimate elements of which the universe is composed, yet they have no independent existence in nature. Their character and behavior can be analyzed mathematically, but nothing else can be said about them. This was a total departure from the ideals of classical physics and mechanics, from the tradition reaching from Newton, the ideal that held that, in principle, the position and path of any and every particle of matter could be known and the structure of the universe therefore charted from their interactions. That knowledge was now shown to be out of reach.

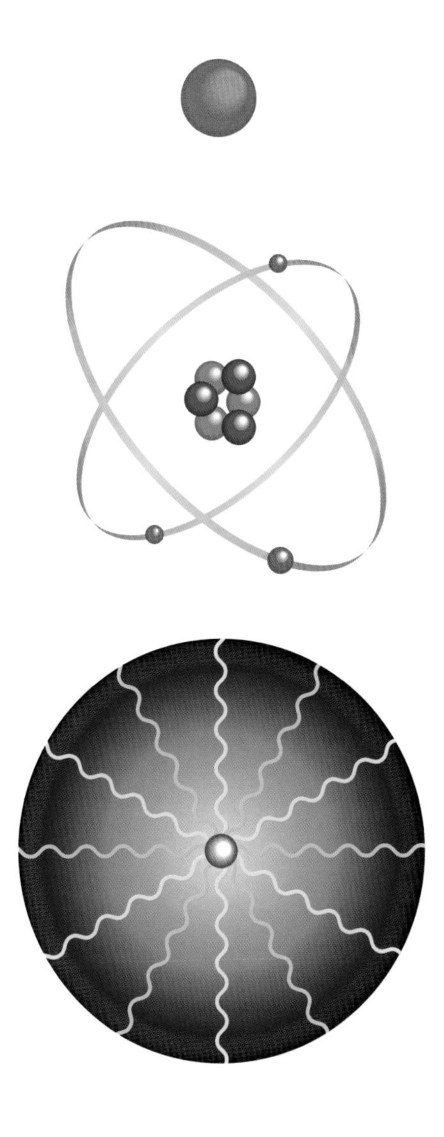

Evolution of the idea of the atom. From top to bottom, the initial, granular view of Democritus; Rutherford's idea of electrons orbiting the neutron; the quantum mechanical version of Schrödinger.

SHIFTING SANDS

The philosophical implications of the new quantum physics were not lost on its supporters. The assertion that the structure of matter was, at its deepest level, fluid, uncertain, incapable of observation or prediction was disturbing. Many were reminded of the ancient Greek idea that nature was a Heraclitean fire (see Volume 1, page 35) where change is the only constant. Men like Planck and Schrödinger were products of conservative, cultured families, and they were reluctant to follow the nihilistic direction (denying the existence of faith and the work of God) in which their work seemed to be leading. They retained their faith that human reason participated in some way in the laws of nature, and that the two should never be in conflict, as quantum physics seem to show they were. The old-fashioned idea of the atom as a solid building block had now shifted until it was seen as an undulating charge of electrical energy, a group of superimposed waves. And from these waves and charges of energy all the material universe is built up—stones, trees, stars, humans, clouds, ships, rivers, everything. All electrons and all protons were identical and featureless, and the only difference between an atom of hydrogen and an atom of uranium was that the first had one of each, while the second had 92 of each. From such differences arose all the range of chemical elements and compounds and all the diversity of natural forms. The science of matter as it developed from Lavoisier to Schrödinger offered on the one hand laws with clarifying and unifying powers, and on the other the insoluble mystery of how such complexity could arise from such simplicity.

The Redefinition of Space and Time: Albert Einstein

Bohr and Einstein, the two most influential physicists of the twentieth century.

In the new physics that took shape in the early twentieth century one scientist stands out because he revolutionized our understanding not of the microscopic nature of matter but of the macroscopic (visible) conditions under which matter forms the universe. In order to describe the mechanics of the universe, three basic factors are required: space, time, and mass. It was Albert Einstein (1879–1955) who showed that these three things are related to each other in ways that had never before been truly understood. Einstein was often pressed for a quick, nontechnical summary of this ideas, and on one occasion he replied by saying that before his theory of relativity, it had always been assumed that if all matter vanished from the universe, space and time would remain; according to relativity, he said, this was not true.

ANALYZING MOTION

Einstein's theories were published between 1905 and 1916. His center of interest was to analyze motion in different frames of reference. A ship on the sea, a passenger on a ship, the sea on the Earth's surface, the Earth itself, the solar system—all were in motion in different directions, and Einstein asked whether there was any absolute standard of time and space by which these movements could be related to each other. His answer was that there is not, because all events are defined by light, and light takes a finite time to travel, so that any event or any movement will appear to occur at a different point in time and space to observers in different frames of reference; there are in fact no simultaneous events.

DEFINING LIGHT

Einstein's theories aroused worldwide interest, partly because they appeared so paradoxical, running counter to common sense; but it should be emphasized that the effects of relativity are not detectable in everyday experience. What Einstein did was to construct mathematical models involving scales of speed and distance far greater than those encountered in normal life and to calculate their effects. In this way he calculated that as speed increases, time moves more slowly, until finally, at the speed of light, time stops altogether. How can this be so? Because light defines all events: If something were moving at the speed of light, light could never catch up, and the event could never happen. This is one of the reasons why the speed of light is the top limiting velocity in the universe. The same effect is true at much lower speeds, those of a train, for example, but they are not detectable by normal means.

In the same way Einstein showed that physical mass increases with speed. In classical physics a body's mass is regarded as constant irrespective of its motion, but Einstein showed that this was untrue. If mass is defined as resistance to change of motion, then it is clear that a greater force is needed to move an object at a greater speed. At very high speed more and more force is required to increase the speed even further, until at theoretical speeds verging on that of light, infinite force would be required for any increase. Once more it was proved that nothing could exceed the speed of light. This aspect of relativity has been verified over and over again at the atomic level by studying the behavior of particles in accelerators. Infinitesimally small electrons at speeds approaching that of light gather enormous mass.

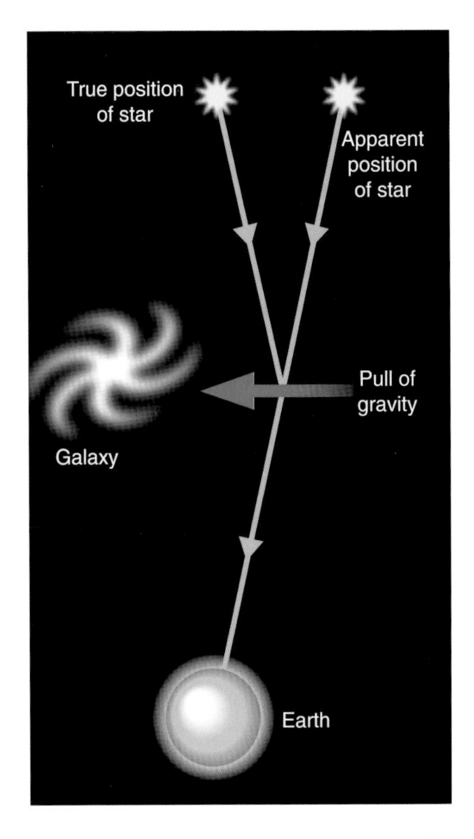

Above: Einstein predicted correctly that light would be bent by gravity, as part of his contention that mass and energy were interchangeable.

ENERGY AND MASS

Following from this discovery, Einstein went on to make a deduction of huge importance both intellectually and practically. Since the mass of a moving body increases with motion, and since motion is a form of energy, the increased mass must come from the energy. Therefore, Einstein argued, energy has mass, or rather, energy and mass are interchangeable, and the distinction is one of temporary states.

By taking his calculations of the effects of increasing speed toward that of light, Einstein arrived at the celebrated formula $E = mc^2$, which showed the amount of energy concentrated into any physical mass: In any particle of matter the energy locked up is equal to the mass, in grams, multiplied by the square of the speed of light, in centimeters per second. This means that if one kilogram of coal were entirely transformed into energy, it would yield 25 billion kilowatt hours of electricity, enough to power a nation for weeks. This transformation could only occur by disintegrating the atomic nuclei and should not be confused with the chemical reaction that occurs in normal burning. But it must be remembered that when Einstein formulated this equation in 1905, neither he nor any one else had any clear idea of the structure of the atom or of the way in which its energy could be unlocked. Einstein was the father of nuclear power and nuclear weapons, but in an intellectual sense only.

Below: Einstein's universe. The universe is composed of matter, space, and time, and Einstein explained how all three are interdependent: without matter there is no absolute space or time, in the same way a room is only defined by the four walls which enclose it. This means that space and time are fluid and have properties which can change under different conditions.

ABIDING MYSTERY

The convertibility of mass and energy provides part of the answer to some of the deepest puzzles in physics. It shows how radioactive substances such as uranium can send out charged particles for thousands of years. It explains how the Sun and other stars can go on radiating heat and light for billions of years. Perhaps above all, it illuminates the dual nature of matter, why sometimes it behaves like charges of radiation or electricity, but at other times it has measurable mass. If matter loses its mass and travels with the speed of light, it becomes radiation or energy. If that energy stabilizes into mass, it forms chemical elements, and we call it matter. Yet the mystery remains: What it "really is," where it originates, and what its ultimate end will be. The whole thrust of science in the nineteenth and twentieth centuries was toward the unification of nature's forms and forces: the conservation of matter, the conservation of energy, the laws of thermodynamics, and now Einstein's equivalence of matter and energy. All these fundamental laws seem to show that the universe is a single process: It is a succession of forms taken on by elements that are themselves eternal, moving in a vast, complex, and unending cycle.

DEPTHS OF THE COSMOS

It is on the atomic and cosmological level that Einstein's ideas have been most profoundly influential. If space and time only exist relative to matter, then it no longer makes sense to separate them: Instead, they form a space-time continuum. That continuum is defined by light, and light, as mass-energy, should be subject to gravity—it should curve around matter. Gravity, in Einstein's view, was therefore the curving of space-time due to the presence of matter in the universe: If there were no mass-energy in the universe, there would be no space-time.

At speeds approaching that of light, time slows and mass increases.

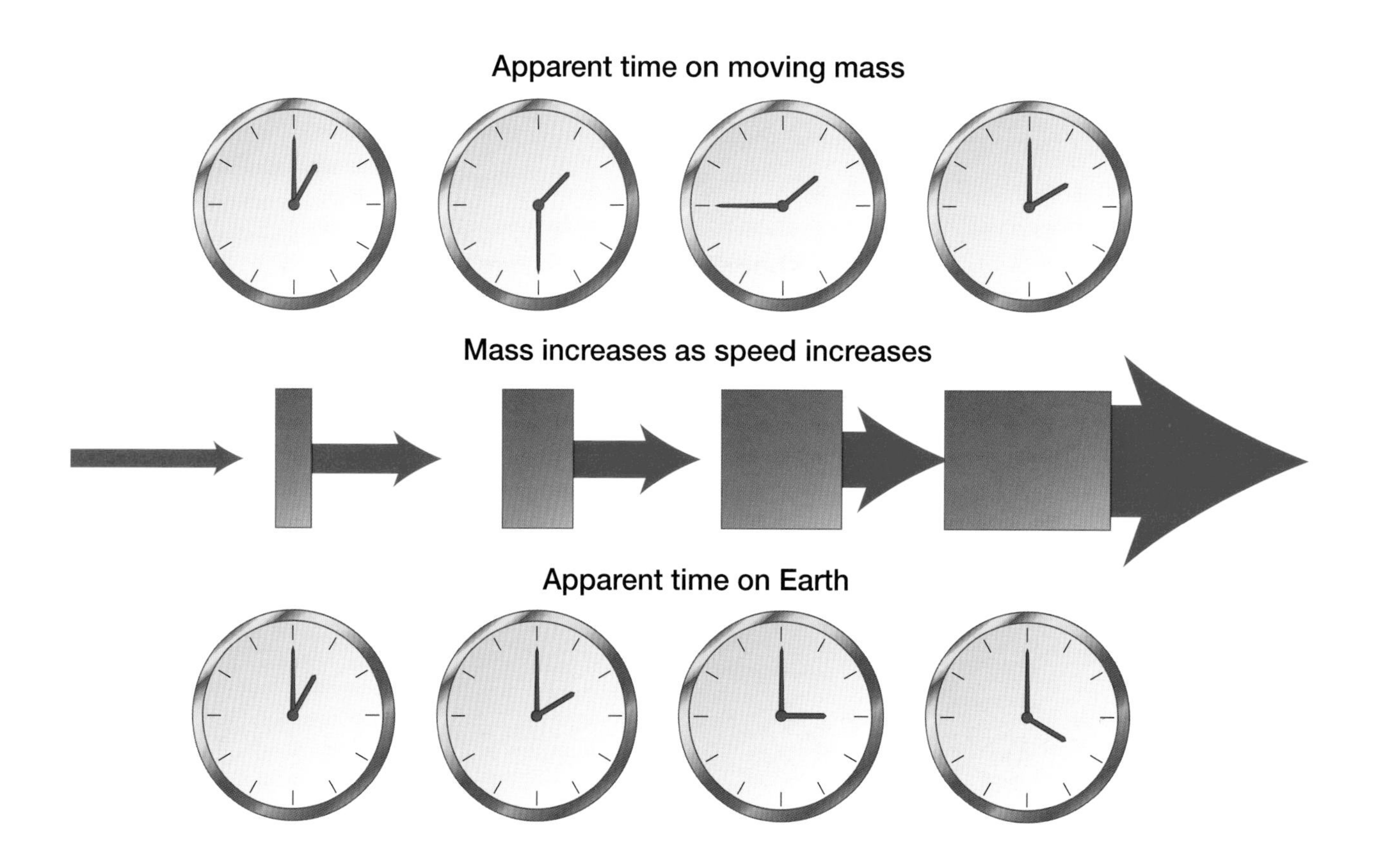

The bending of light in the gravity field of a massive star—the Sun—was verified in a famous experiment carried out in 1919 by the English astronomer Sir Arthur Eddington (1882–1994). This concrete proof of Einstein's bizarre and apparently abstract theories resulted in his being enveloped in a tidal wave of fame and hailed as the author of a new vision of the universe. The very strangeness of his ideas added to his celebrity, for example, when he announced that the universe must be considered finite but without limits. This concept was also proved true by the new cosmology that emerged between 1920 and 1950.

It was commonly said that Newton's physics had been overthrown, but Newton's physics still governs our terrestrial experience—for example, they are all that is needed for aerial navigation or for space travel within the solar system. Einstein's physics take us into the realms of the atom, the speed of light, and the depths of the cosmos.

The last 30 years of Einstein's life were spent in the search for a "unified field theory" that would embrace gravity and electromagnetism, the cosmos, and the atom. He was deeply distrustful of quantum physics and felt that it must be possible to build a model of physical reality as it really is and a mere probability system. His criticism of quantum theory—that "God does not play dice with the universe"—has become famous. Perhaps Niels Bohr's reply deserves equal weight, that "we cannot tell God how to organize the universe."

ALBERT EINSTEIN (1879–1955)

- Mathematical physicist.
- Born Ulm, Germany.
- Educated at Munich, Aarau, and Zurich
- 1901 Took Swiss nationality.
- 1902–5 Worked at the Swiss Patent Office.
- Published papers on problems in theoretical physics.
- 1905 Published his theories on relativity, which won him worldwide acclaim.
- 1909 A special professorship was created for him at Zurich.
- 1914–33 Director of the Kaiser Wilhelm Physical Institute, Berlin.
- 1916 Published further theories on relativity.
- 1921 Won the Nobel Prize for physics.
- Left Germany when Hitler came to power and moved to Princeton in 1934.
- September 1939 wrote to President Roosevelt about the potential of an atomic bomb.
- 1940 Took U.S. citizenship and became a professor at Princeton University
- After World War II agitated for international atomic weapons limitation and control.

Einstein with his daughter and son-in-law.

The Manhattan Project:
Science and Secular Power

The new physics from Rutherford to Heisenberg represented an intellectual revolution in our understanding of the physical universe. But there was one episode that carried this revolution far beyond the laboratory: The Manhattan Project, the code name for the building of the first atomic bomb, involved the leading physicists of the world and their newest theories, and it raised questions about the aims and results of scientific progress that have grown more and more pressing as the years pass. The atomic bomb was designed and built by Allied scientists between 1941 and 1945 in response to the fear that Nazi Germany might develop the weapon first. The first of the ironies surrounding the Manhattan Project is that Nazi policies had driven out of Germany many of the key scientists whose ideas and skills made the bomb possible, either because they were Jewish, or because they opposed Nazism.

The Italian physicist Enrico Fermi.

It was in Italy in the late 1930s that Enrico Fermi (1901–1954) discovered that when certain elements were bombarded with low-velocity neutrons, their atomic nature was changed, and they formed different elements. This result was studied and analyzed by scientists in many countries, who figured out that the nuclei were splitting, and that high levels of energy were released in the process.

CHAIN REACTION

In 1939 no less a figure than Niels Bohr, who was then working temporarily in America, discovered that in this fission process additional neutrons were released. If the proper amount of material

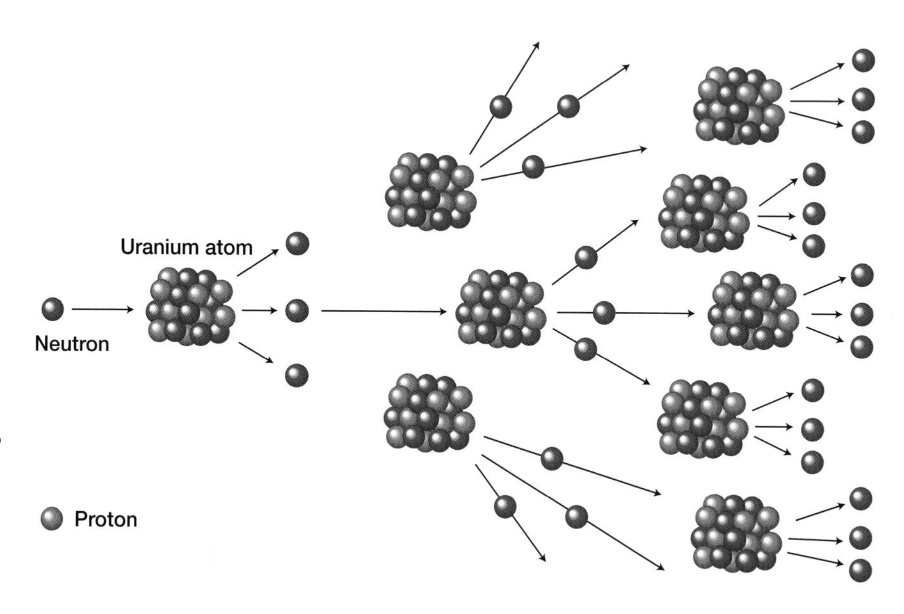

Chain reaction: Each uranium atom as it splits releases neutrons, which go on to split more uranium: the huge energy that binds the atoms is released in an instant.

were present under the right conditions, it seemed that a very fast chain reaction of nucleus splitting might be started: Each neutron would split a nucleus, freeing more neutrons to split more nuclei, and the process would explode, releasing huge levels of energy. How this could be done was unknown, and its application was unclear, except to the obvious one of making a weapon of unprecedented power. This kind of experiment and its possible results were certainly known to physicists in Germany, and Bohr felt it essential to warn of the danger. In the summer of 1939, together with the other European refugee-scientists Edward Teller (1908–) and Leo Szilard (1898–1964), he persuaded Einstein, the most famous scientist of his day, to write a letter to President Roosevelt urging "watchfulness and, if necessary, swift action" in nuclear fission research. The United States was not yet at war, but an advisory scientific committee was indeed set up, and basic research was started at Columbia University. In March 1940 it was confirmed that nuclear fission using uranium or plutonium was a strong theoretical possibility, but that the design and control of the process in weapon form would require years of work.

Albert Einstein
Old Grove Rd.
Nassau Point
Peconic, Long Island

August 2nd, 1939

F.D. Roosevelt
President of the United States,
White House
Washington, D.C.

Sir:

Some recent work by E. Fermi and L. Szilard, which has been communicated to me in manuscript, leads me to expect that the element uranium may be turned into a new and important source of energy in the immediate future. Certain aspects of the situation which has arisen seem to call for watchfulness and, if necessary, quick action on the part of the Administration. I believe therefore that it is my duty to bring to your attention the following facts and recommendations:

In the course of the last four months it has been made probable - through the work of Joliot in France as well as Fermi and Szilard in America - that it may become possible to set up a chain reaction in a large mass of uranium, by which vast amounts of power and large quantities of new radium-like elements would be generated. Now it appears almost certain that this could be achieved in the immediate future.

This new phenomenon would also lead to the construction of bombs, and it is conceivable - though much less certain - that extremely powerful bombs of a new type may thus be constructed. A single bomb of this type, carried by boat and exploded in a port, might well destroy the whole port together with some of the surrounding territory. However, such bombs might very well prove to be too heavy for transportation by air.

The historic letter sent by Einstein to President Roosevelt, which initiated serious work on atomic fission. Einstein blamed himself ever afterward for writing it.

CATASTROPHIC POWER

America's entry into the war in December 1941 put the project on a much more urgent footing. Huge funds were made available, and hundreds of leading scientists were recruited, many of them refugees from Europe: Enrico Fermi, Edward Teller, Bohr himself, Hans Bethe, and John von Neumann, while many American scientists joined who would later achieve fame in other fields, such as Richard Feynman (1918–88). The problems were both

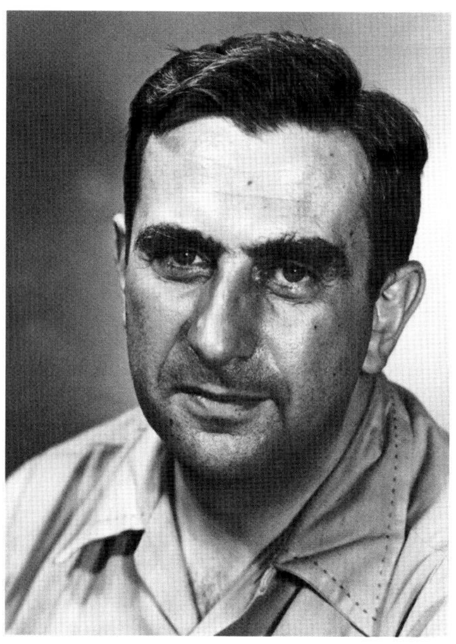

Top: J. Robert Oppenheimer, the brilliant young research director of the Manhattan Project.

Above: Edward Teller went on from the Manhattan Project to work on the first hydrogen bomb.

theoretical—to predict what would actually happen in the chain reaction—and practical—to design a procedure to make it happen at the required moment. This was like no other engineering project ever undertaken, for it involved fundamental research into the nature of matter, and its aim was to unleash an event of catastrophic destructive power.

Every stage of the project was a venture into the unknown. Even the basic material of the weapon was uncertain, for neither uranium nor plutonium occur in a pure form in nature, and their extraction is complex and costly. Never had new theoretical knowledge been converted so quickly into new technical systems. A major milestone came in December 1942, when Fermi produced the first controlled fission chain reaction in a Chicago laboratory. But the physical design of a bomb would be very different from an apparatus to be used in a laboratory, and a thousand problems had to be overcome in containing the fissionable material and in detonating it at the required time.

In June 1942 the project was placed under the direction of J. Robert Oppenheimer (1904–67), a brilliant young physicist whose task was to coordinate the different work groups, now involving thousands of scientists and technicians, and centering on the Los Alamos site in the New Mexico desert. As the war entered 1945, Germany's defeat became inevitable, and it appeared that the weapon's final target would be Japan.

It is a matter of record that all the scientific and technical problems were solved, and that a test bomb was exploded in the desert on July 16, 1945. From a few ounces of plutonium it yielded the equivalent of 21,000 tons of TNT, vaporizing the steel tower where the bomb was mounted and fusing the desert sand to glass within a radius of half a mile. Into Oppenheimer's mind came some words from an ancient Hindu poem, "I am become death, the shatterer of worlds." Within three weeks two bombs had been

detonated over Japan, ending the war, but with the deaths of some 200,000 inhabitants of the cities of Hiroshima and Nagasaki.

"I am become death, the shatterer of worlds." The words of an ancient Hindu poem that occurred to Oppenheimer when he contemplated the destructive power of the atomic bomb.

NATURE'S INMOST FORCE

Some of the leaders of the Manhattan project, notably Edward Teller, went on to develop a second generation of nuclear weapons working on the different principle of the fusion of atomic nuclei, which releases even greater levels of energy—the so-called hydrogen bomb. But most of the figures involved clearly felt a mixture of pride and shame at what they had done. They had pursued scientific theories to their ultimate ends, seeking knowledge of nature's inmost forces, but the results had been a catastrophic loss of human life. Of course, it was possible to justify this project in the context of the war, but the true feelings of these scientists was shown when most them chose to devote their later lives to

No ordinary tennis party, these were the physicists engaged in building the atomic bomb in 1946.

ROBERT OPPENHEIMER (1904–67)
- Nuclear physicist.
- Born New York City, New York.
- Attended Harvard, Cambridge University, Leyden, Zurich, and Göttingen universities.
- 1927 Received his doctorate from Göttingen University before returning to the United States.
- Founded the school of theoretical physics at Berkeley and the California Institute of Technology.
- Studied and researched nuclear physics, in particular cosmic-ray theory, electron-positron pairs, and deuteron reactions.
- 1943–45 Set up and ran the Los Alamos laboratory to research and build the atomic bomb.
- 1947 Director of the Institute for Advanced Study at Princeton University.
- 1946–52 Chairman of the advisory committee to the U.S. Atomic Energy Commission.
- Openly and vociferously opposed the development of the hydrogen bomb and was removed from secret nuclear research.
- 1953 Declared a security risk and forced to withdraw from the public eye.

preventing the use of nuclear weapons. Einstein in particular was horrified at the results of his original insight into the equivalence of mass and energy. Moreover, after the war's end it became known that German efforts at nuclear-weapon research had never passed beyond the most elementary stage. Those physicists who had remained in Germany, such as Heisenberg, had apparently made fundamental miscalculations about many important matters, such as the amount of uranium that would be required. In fact, not a fraction of the effort had been devoted to the work in Germany as had been invested in the American enterprise, and a German atomic bomb had never come anywhere near being built. This was the second great irony of the Manhattan Project.

The invention of the atom bomb marked the emergence of science into the arena of political and moral consciousness, for it had been the product of pure theoretical science, the work of the world's leading professional scientists, and it had resulted in horrifying destruction. Moreover, it was not a finite event, for it inaugurated an age in which international politics would be dominated by the threat posed by such weapons. The real questions raised by the Manhattan Project are these: Should scientific truth be pursued at all costs? Can the scientist detach himself from the consequences of his work? Do scientists have the right to reshape the world in which the human race must live? Are there times when the truth is best left hidden? More recently, the urgency of these questions has shifted from weapons to the field of biological research.

Geophysics: The Dynamic Earth

By the early twentieth century the traditional study of geology—classifying rocks and landforms—had gradually broadened into the more inclusive science of geophysics, which sought to understand the structure of the Earth as a whole: the nature of the Earth's core, how land and sea interact, the true origin of volcanic activity, the cause of the Earth's magnetic field, and other general questions like these.

One of the first breakthroughs of the century concerned the age of the Earth. The theories of Lyell and Darwin (see Volume 8, pages 19–25) had demanded that geological time must be measured in hundreds of millions of years to allow for the accumulation of changes in both living and nonliving matter. Leading physicists had pointed out that this required an age for the Earth that was impossible, given the current belief that the Earth had been steadily cooling since its formation. Twenty million years was suggested as a credible age for the

Wegener's original conception of continental drift, first published in 1915, but not widely accepted for many years.

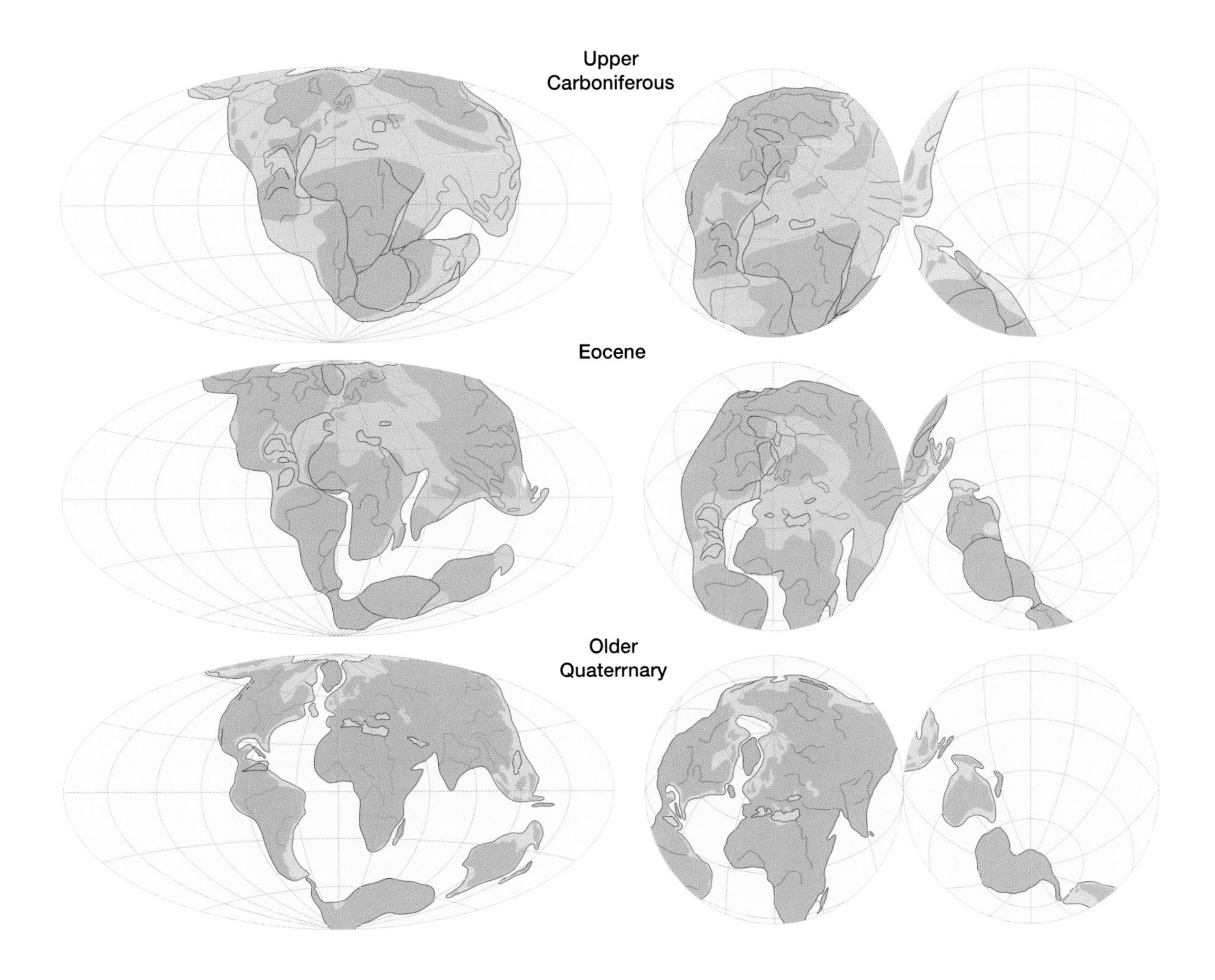

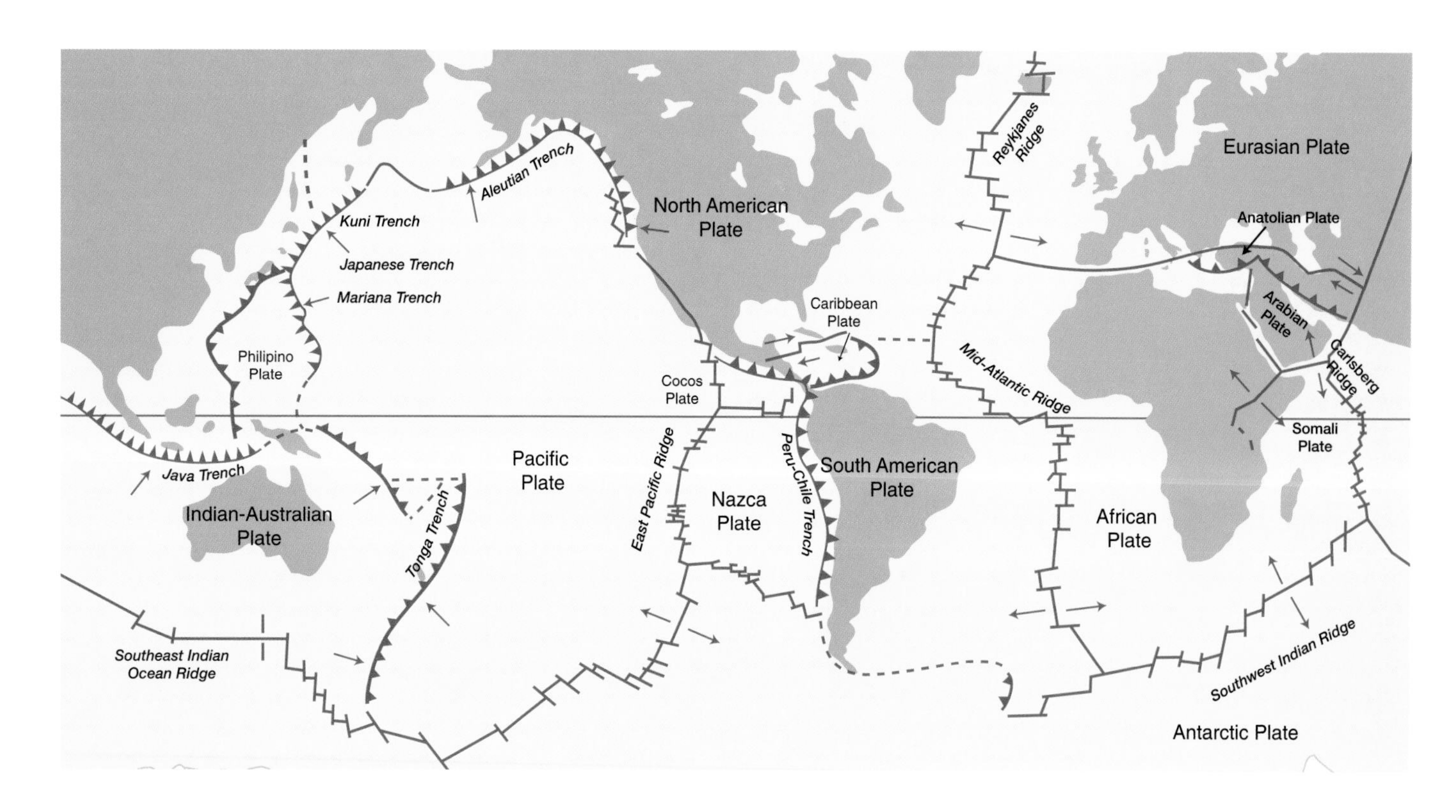

Above: These lines show where the continents ride on their massive plates.

Opposite: Aerial photograph of the San Andreas fault as it crosses the Carrizo Plain some 300 miles south of San Francisco and 100 miles north of Los Angeles. The fault, which extends almost the full length of California and is responsible for major earthquakes, is the narrow, valleylike scar running from bottom right to top center of the picture. The desolate landscape consists of pressure ridges formed as a result of hundreds of fault movements. The Carrizo Plain is at left, the elevated Elkhorn Scarp at right. This is the place where the San Andreas fault is at its most spectacular.

planet. But the discovery of radioactivity as a long-term source of heat within the Earth altered this picture completely, and by 1910 the American chemist Bertram Boltwood (1870–1927) had used the rate of radioactive decay to calculate that the Earth could be as much as two billion years old. This was steadily extended until estimates in the 1970s reached 4.5 billion years, a figure that was confirmed from meteorite samples and lunar rocks, and that is now accepted as the age of our solar system.

STRIKING FIT

In understanding the processes shaping the Earth's crust, the theory of plate tectonics has revolutionized virtually every aspect of geophysics. It explains seismic activity, volcanoes, mountain building, and the character of the ocean bed.

The first step toward the theory of plate tectonics came from the German meteorologist Alfred Wegener (1880–1930). Wegener's starting point was the striking fit between the coastlines of some of the continents, most obviously between the east of South America and the west of Africa. It had been noticed before by geographers, but Wegener believed that it must be more than a coincidence. In his book *The Origin of Continents and Oceans,* published in 1915, he proposed that throughout most of geological time the Earth had contained only one huge landmass, which he christened "Pangaea" (Greek for "whole earth"). Around 200 million years ago, he suggested, it had begun to break up into the separate continents that we know today, and they had drifted across the surface of the globe.

Wegener produced evidence for this startling theory by showing that the rock types along opposite continental coasts often matched each

Thermal forces in the magma, the molten layer under the Earth's crust, make the tectonic plates move. Sometimes, at the edges of the plates, the magma wells to the surface through volcanoes—as with this lava fountain during a volcanic eruption at night. Lava is molten rock that has seeped up from the Earth's outer core (mantle) through cracks in our planet's crust. The lava is forced out because of the high underground pressures. Some of the lava constituents become gaseous as the pressure reduces; these gases force out the lava explosively to form fountains. Photographed on the island of Bali, Indonesia.

other very closely, and it also helped to explain why many species of animals had developed that were unique to one continent, such as the marsupials of Australia, while others were common to all continents. Wegener's theory met a cool response, with many scientists simply refusing to believe that something as massive as a continent could drift. Wegener died in 1930 during an expedition to Greenland before he could see his ideas accepted. A different version of Wegener's theory was produced in the 1930s by the South African geologist Alexander du Toit (1878–1949), who argued that there were two supercontinents, "Laurasia" in the north and "Gondwanaland" in the south.

MAGNETIC PATTERNS

There were strong arguments in favor of continental drift, but no one could conceive any mechanism to drive it. Further evidence emerged in the 1950s, first from the magnetic patterns found in certain ferrous (iron-containing) rocks. Studies of the magnetism in these rocks seemed to show that when they were formed, the Earth's magnetic poles were in quite different positions from their modern ones. Evidence built up from rocks from all the continents strongly suggested that they were once joined together, just as Wegener had claimed. Another major breakthrough came with the postwar generation of seabed surveys, which showed that all the ocean floors were made of much shallower rock than the continents and were also much younger. The ages and magnetic patterns of the seafloor rocks made symmetrical patterns on either side of the ridges that were to be found in the middle of all the ocean beds.

It was a Canadian geologist, John Tuzo Wilson (1908–93), who drew these threads together and announced the startling theory that the surface of the Earth is divided into half a dozen large plates, with some smaller ones, that ride on the molten layer beneath the Earth's crust. Hot new rock is working its way to the surface in the midocean ridges, while where the edges of the plates meet, they grind against each other and produce earthquakes, for example, along North America's West Coast. Major mountain building, for example, the Himalayas region, was caused when one plate crashed into another. The energy for the movement of the plates is believed to come from thermal forces in the magma, the hot, molten layer that lies beneath the Earth's crust.

Modern opinion is that the continental plates have existed in their present form for at least 500 million years, and that they did indeed come together to form a "Pangaea" that then began to split apart some 200 million years ago, as Wegener had guessed. Plate tectonics is a powerful and well-supported theory, but it is nevertheless mysterious. Just as the other modern physical sciences seem to have shaken the firm foundations of our knowledge of matter and the universe, so geophysics has shown us that even the solid earth beneath our feet is moving in obedience to the immense forces of nature.

The Structure of the Universe: Evolution among the Stars

Twentieth-century astronomy has given birth to a set of radically new ideas about the structure of the universe. These ideas took shape in a series of steps in which new observations and new techniques of interpreting them showed scientists how they might understand the scale of the universe and even perhaps how it had come into being. The findings of chemistry and physics were increasingly used to build up models of the processes at work in stars and between stars, thus creating the new science of astrophysics.

STAR CATALOGUE

Spectroscopy—the study of the emissions of light and energy given off by substances—was the first clue. At Harvard in the 1890s a team of astronomers led by Edward Pickering (1846–1919) was engaged in a massive program of analyzing the spectra (light waves) of thousands of stars. One of the astronomers, Annie Cannon (1863–1941), noticed that these spectra fell naturally into around ten types, from those in which the blue light predominated, to those where the red was dominant, and in which the dark lines that showed the presence of the various elements fell into patterns. In 1901 a catalogue of over a thousand such spectra was published, and 20 years later the Harvard star catalogue had grown to number over 225,000, all confirming the same pattern.

Many astrophysicists saw at once that these spectra revealed the surface temperature of the stars, and two astronomers, working independently, set out to relate these temperature spectra to the visual magnitudes, (the brightness) of the stars. The Dane Ejnar Hertzsprung (1873–1967) and the American Henry Russell (1877–1957) both produced diagrams showing that the great majority of stars were grouped into one main sequence moving from high temperature and high luminosity (light) to much lower levels of both. Further analytical techniques enabled them to calculate the relative sizes of many stars, and this produced two important subgroups that broke the predominant pattern: One group of stars was very hot, but was far less bright than they should be because they are very small, while another group was low-temperature but very bright because they are exceptionally large.

The Hertzsprung-Russell diagram, which relates star temperature to luminosity, from low levels in the bottom right to high levels at the top left.

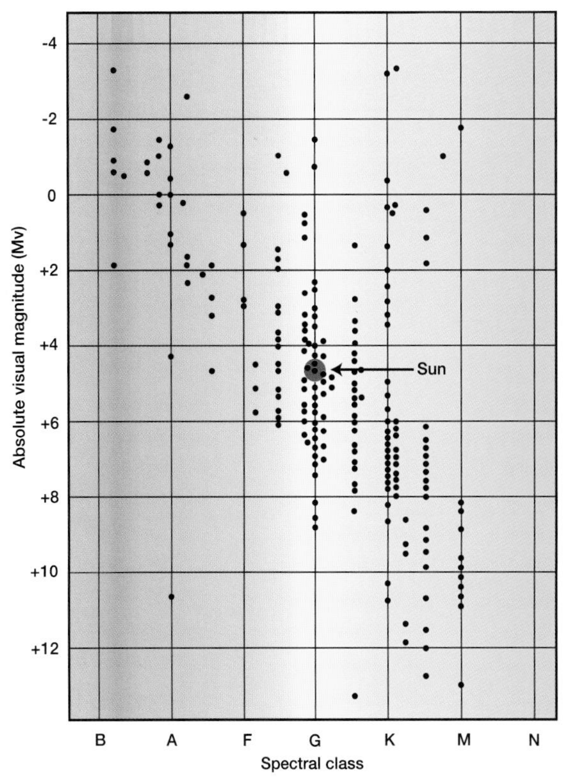

STAR LIFE CYCLE

What did all this mean? Did it mean that there existed a number of fundamentally different types of stars functioning in different ways? Both Hertzsprung and Russell took a different view, for they concluded that that these spectra showed that stars were linked in a process of evolution: What they were seeing in each case was a star at one particular stage in a cycle through which all stars would pass. But which way was the cycle moving? Russell thought that stars began as large, red, relatively cool bodies and grew steadily hotter and denser, while Hertzsprung took the opposite and correct view that stars were gradually cooling.

The Hertzsprung-Russell diagram first appeared in 1913 and quickly became one of the astrophysicist's fundamental tools, showing the place of any star in the evolutionary tree of the cosmos. The classical belief that the heavens were eternally unchanging had long ago been abandoned, but the possibility that the stars themselves were dynamic—they were, to put it simply, living and dying—was a startling new idea.

ATOMIC FUSION

But what was actually happening as the stars burned hotter or cooler? The fire in the Sun and stars that could burn for thousands, or perhaps even millions, of years without being consumed had long been a mystery to astronomers. Before the advent of atomic physics no known mechanism could account for the stability of the Sun. It was the British astrophysicist Sir Arthur Eddington (1882–1944) who applied atomic theories to the stars. Eddington used Einstein's discovery of the equivalence of mass and energy to suggest that the source of solar energy is atomic fusion, that hydrogen atoms were fusing to helium, releasing huge amounts of energy. In his classic book *The Internal Constitution of the Stars* (1926) he also calculated the mass of the Sun and suggested a lifetime of billions of years, which was in line with the processes embodied in the Hertzsprung-Russell diagram. As Rutherford's work on the atom had opened up new ideas on the age of the Earth, Eddington applied the same principle to the Sun and stars.

Eddington was not able to put forward a precise model of the atomic transformations occurring inside a star, and it was some years later that Hans Bethe (1906–), one of the many refugee-scientists from Germany working in America, gave a more detailed answer. Bethe showed that six separate atomic transformations were required, leading from hydrogen to helium, with carbon acting as the vital catalyst. Further detailed studies would show that the hottest stars were the youngest, and that they ended their life as the red giants and white dwarfs that formed the important subgroups in the Hertzsprung-Russell diagram. Eddington and Bethe had shown that the structure and behavior of the atom, so bizarre, unexpected, and powerful, turned out to hold the key to large-scale cosmic processes too.

Opposite: The colossal energy of the Sun springs from nuclear fusion and will continue for billions of years.

Below: Sir Arthur Eddington, who proposed that the stars are fueled by atomic energy.

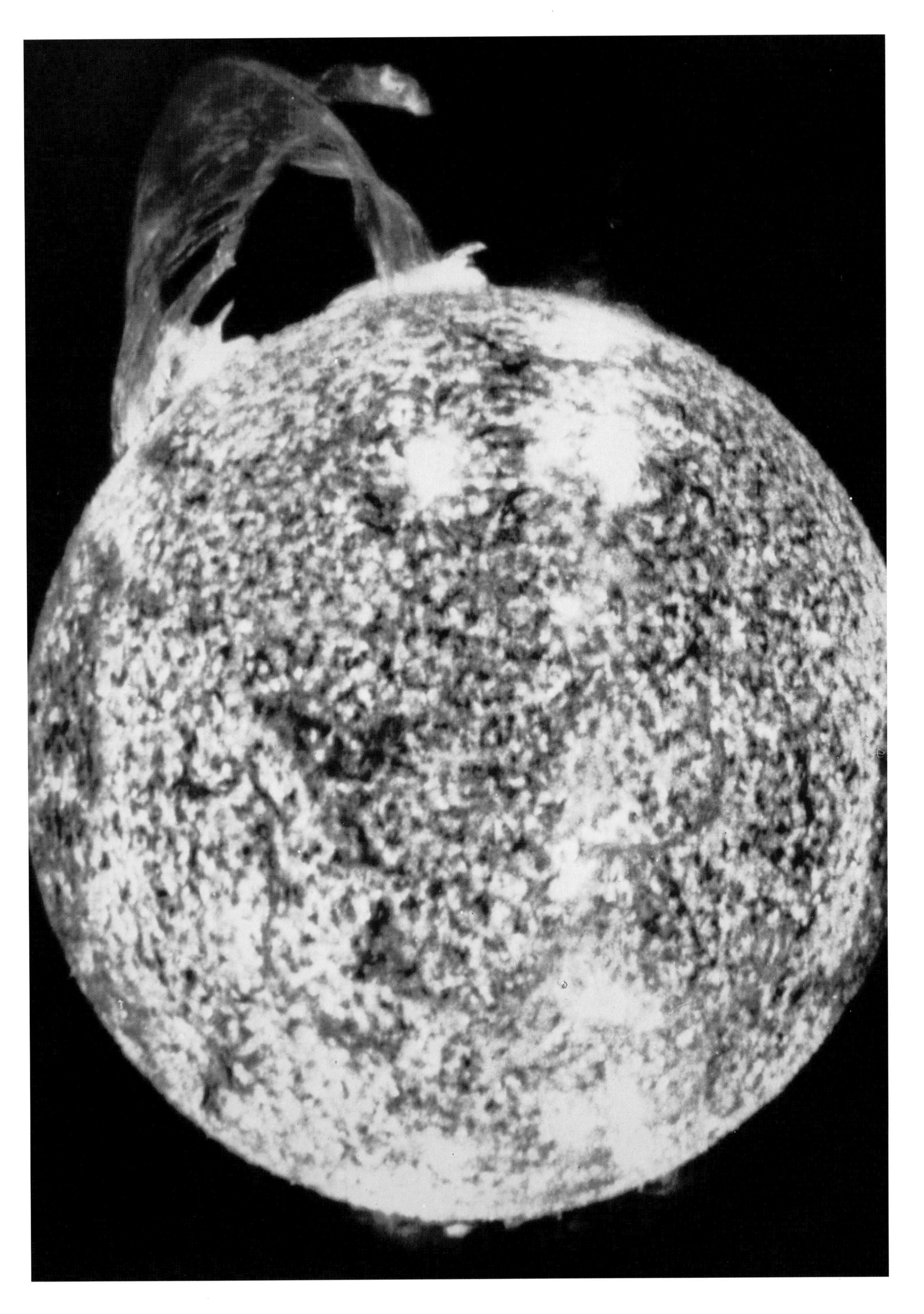

To Infinity and Beyond: The Scale of the Universe

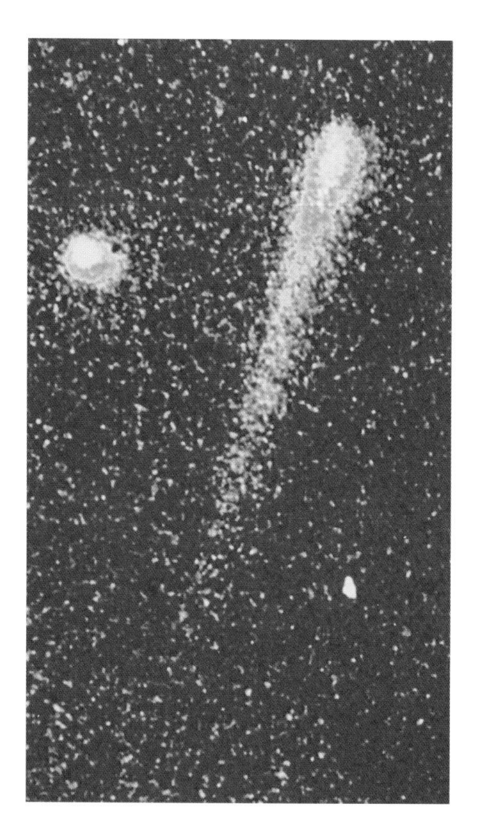

December 25, 1974, saw the comet Kohoutek visible in the skies.

Before humankind could understand the structure of the universe, it faced the fundamental problem of grasping its scale. All the thousands of objects that filled the sky, and the millions more that were being revealed by powerful new telescopes, lay scattered through space at vastly different distances from the Earth. But were all these objects part of one system, or were there distinct groupings or levels among the heavenly bodies? This was the old problem that Herschel had first addressed: How to map the heavens in three dimensions, a far more difficult task than simply locating points on the celestial sphere. For the nearer stars the established technique of parallax calculation (see Volume 6, page 52) could give a reasonable idea of stellar distances, but for the more distant stars some quite different measuring guide was needed.

VARIABLE STARS

Such a guide was discovered in 1912 by Henrietta Leavitt (1868–1921), who was another astronomer working on the Harvard star catalogue under Edward Pickering. Leavitt was studying a group of variable stars—stars whose brightness rises and falls over a period usually of several days. In the stars she was studying, which were all located in the Small Megellanic Cloud deep in the southern sky, she noticed that the longer the period over which this change took place, the brighter the star was. This relationship turned out to be quite precise: They were like preset beacons, so that just by observing its period, one could tell how bright it was. Therefore, if such a variable star were seen to have a very long period, but it appeared very faint, this difference between its apparent brightness and its true magnitude would indicate how distant it was.

This new technique was taken up by several astronomers in an attempt to rescale the distribution of stars in the heavens. The most prominent was the young American Harlow Shapley (1885–1972), who was on the staff at Mount Wilson Observatory using the 60-inch reflector, then the largest telescope in the world (the 100-inch was not completed until 1918). By studying examples of these variables at the farthest limits of visibility, Shapley was ready in 1917 to announce his finding that the Milky Way Galaxy must have a diameter of approximately 300,000 light-years (in one light-year light travels a distance of about 6 million million miles).

In its time this was a staggering figure, ten times larger than that suggested by any other astronomer. He considered that the galaxy was shaped like a flattish disk with a bulging center, and that the Sun and our solar system were located near the edge. Shapley's conception of the scale of our galaxy was broadly correct, but it had one unfortunate result: Its size was so immense that Shapley considered that the galaxy was identical with the universe, that even the faintest objects visible in the heavens were contained within this grouping, and that he had therefore solved the fundamental problem of the size and structure of the universe.

STAR SYSTEMS

Most astronomers were reluctant to accept Shapley's view because its scale was so revolutionary, but there were a few who disagreed for other reasons. Among them was Heber Curtis (1872-1942) of the Lick Observatory in California, who had been making a special study of the spiral nebulas. Curtis felt instinctively that these nebulas (vast indistinct clouds of stars) were self-contained star systems outside our own galaxy, but the difficulty was to find some distance marker to settle the problem. Between 1915 and 1920 Curtis was able to identify a number of novas—exploding stars that are among the brightest objects in the heavens—in some of these nebulas. This established in the first place that the nebulas were definitely composed of stars, while a comparison of their very faint visibility with the tremendous brightness of novas that were already known showed that they must be at least a hundred times more distant than any novas ever recorded. Using the brightness of such a nova in the Andromeda nebula, Curtis estimated that it lay 500,000 light-years away, further than even than Shapley would allow, but with the crucial difference that this nebula was an "island universe" in its own right, another star system comparable to our own galaxy.

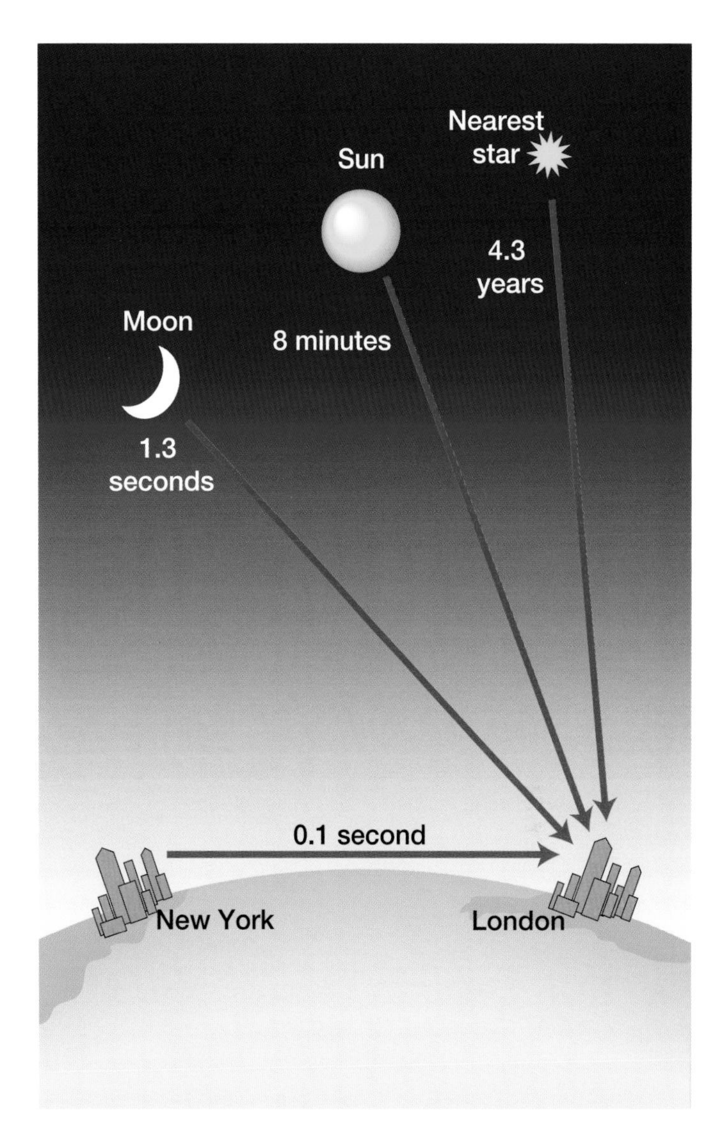

The huge scale of astronomical distances necessitated a new unit, the light-year, which is 186,000 x 31,536,000—the speed of light per second multiplied by the number of seconds in a year. The times taken by light to travel certain distances are shown above.

Curtis and Shapley met in a public debate in April 1920 at the National Academy of Sciences in Washington to present their differing views. This meeting has become famous in the history of astronomy and is often referred to as "The Great Debate," but in fact it was scholarly and inconclusive. Both men were only partly right—Shapley on the immensity of our galaxy, Curtis on the separate identity of the nebulas such as the Andromeda—but they had succeeded in defining the major problem that cosmologists would debate for the next decade.

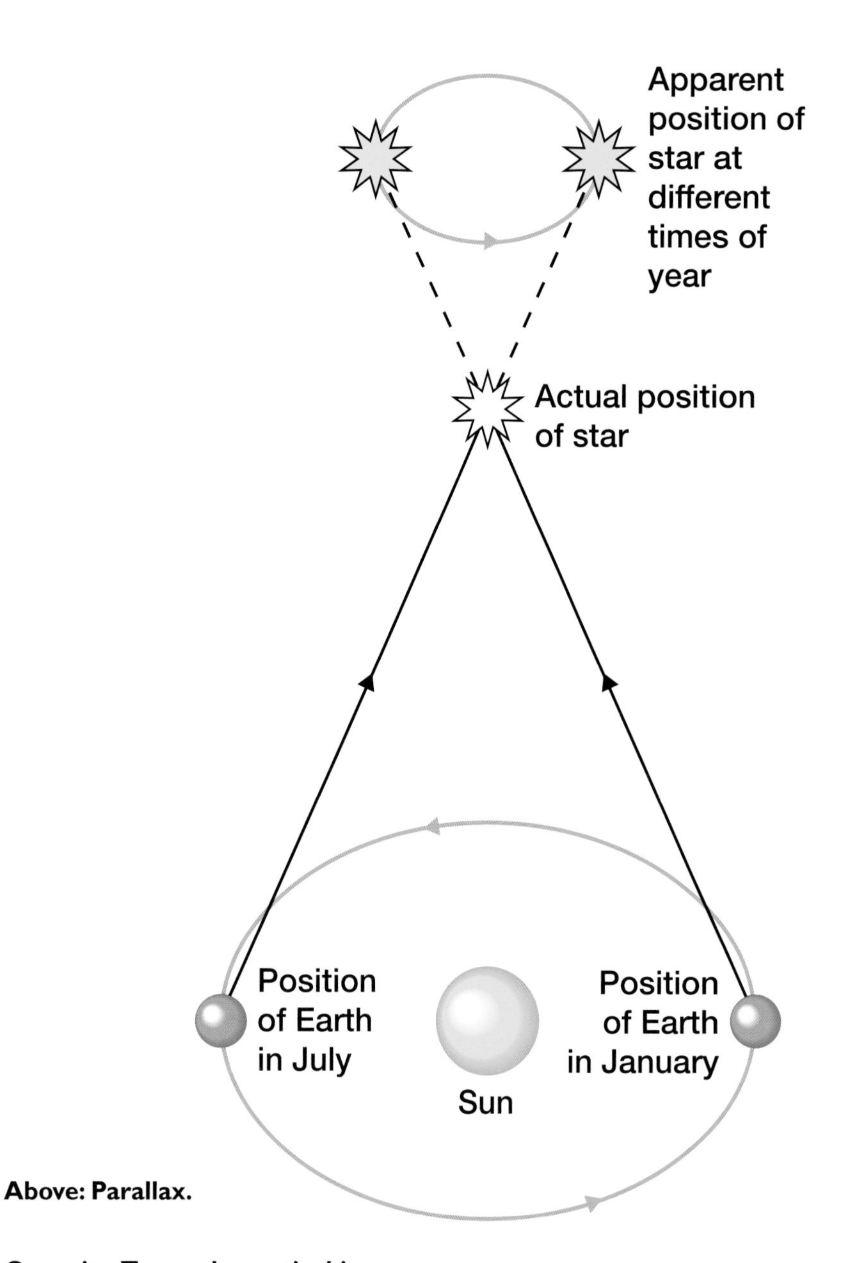

Above: Parallax.

Opposite: True color optical image of the great Orion nebula (M42 and M43, pink, with NGC 1977 above), showing the fine strands of nebulosity in its outer reaches. These nebulas lie around 1,500 light-years away in the constellation Orion. M42 and M43 are HII regions, areas of gas (mainly hydrogen) and dust where newly born stars excite the gas, making it glow. M42 and M43 are excited by a group of stars (the Trapezium) embedded in them. The stars are here obscured by the bright central region. NGC 1977 is a reflection nebula, which reflects the light of stars embedded in it. These typically appear blue since blue light is scattered more than red.

This problem of the scale of the universe was developed on a technical level, using new and sophisticated techniques of observation and measurement. But more than that, it was also an intellectual quest, an attempt by humankind to remap its universe and to try to place itself within it. Just as the atomic physicists of this period were revealing the unbelievable complexities of matter's material components, so cosmologists were grappling with a new vision of the unsuspected scale of the universe. It was a very exciting time to be an astronomer, and the final picture that emerged would exceed anybody's anticipations.

PARALLAX AND TRIANGLES

The traditional method for calculating astronomical distances is the familiar one of calculating triangles from the knowledge of one side's length and two angles. If an observer sees the Moon overhead while another observer, standing on the first observer's horizon, measures his angle of sight to the Moon, then the length of the other two sides—which will be the distance to the Moon—can be calculated. This technique works well for objects that are near Earth, but for distant stars a much longer baseline is needed. Astronomers learned how to use the orbit of the Earth around the Sun as that baseline: Taking observations of a star six months apart would give a baseline that was the diameter of the Earth's orbit, around 190,000 miles. The apparent shift in the star's position was called a parallax shift, exactly the same as the way your outstretched finger seems to jump as you close first one eye then the other. This technique reached its limit when the baseline became so thin in relation to the great distances of the stars that other techniques had to be found.

The Expanding Universe: Edwin Hubble

The impasse concerning the scale of the universe was resolved during the 1920s through the work of another astronomer who arrived at the Mount Wilson Observatory in 1919, Edwin Hubble (1889–1953). Using the new 100-inch reflecting telescope, Hubble looked for a way to settle the conflicting views of Shapley and Curtis, whether our galaxy was coterminous—having the same boundaries in time and space—with the universe, or whether many other similar star systems existed throughout an even wider space.

Edwin Hubble, pioneer of the new age in cosmology.

LANDMARK IN COSMOLOGY

The means to Hubble's decisive new discovery came through the same type of variable stars that Shapley and others had already studied. By 1923 Hubble had identified no fewer than 36 such stars in the Andromeda nebula, and he carefully calculated their periods and their luminosities, from which he estimated that they lay at least 100 million light-years away from Earth. This was far beyond Shapley's largest estimate for the dimensions of our galaxy and was in line with Curtis's ideas on the possible extent of the universe. Hubble's results were announced in December 1924, and it was generally recognized that a landmark in cosmology had been passed, and that a new scale and possibly a new structure for the universe had been revealed. The great question now was whether it was possible to determine how far the extragalactic universe

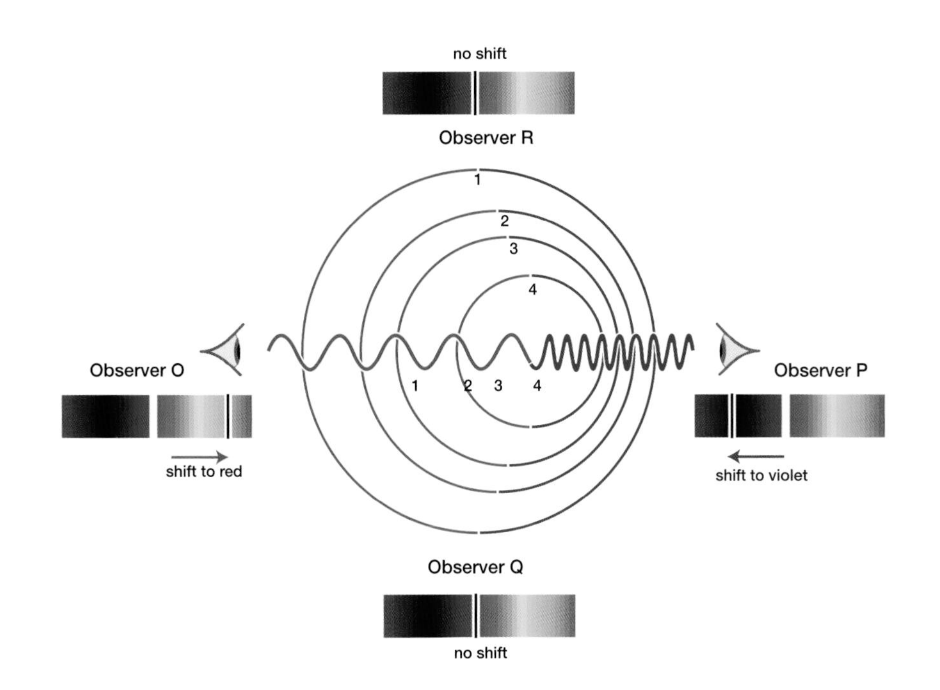

The Doppler effect: Light moving toward the viewer is squeezed into a shorter wavelength; light moving away is stretched into a longer wavelength. The red-shift is a crucial piece of evidence in our picture of the expanding universe.

extended and whether it could be studied. The way forward clearly lay through a study of the galaxies themselves, their character and their distribution in space. How important were the different shapes that the galaxies displayed? Were they distributed at random through space, or was there some significant large-scale structure to the universe?

Hubble played a leading role in this field, and over the following three decades he produced a classification system of galaxies showing the many variations on the basic elliptical and spiral forms, although he did not theorize about any possible process of evolution that might have led to these forms. His other fundamental task was to build up a scale of distances using natural markers such as the variable stars wherever possible.

By comparing types of stars in the distant galaxies, he gradually built up an ever-growing picture of the cosmos, so that by 1930 his observations led him to believe that the visible objects that we see in the heavens extended over a distance of at least 250 million light-years. Hubble was aware that his estimates carried a degree of risk and uncertainty; but when the 200-inch Mount Palomar telescope was commissioned in 1949, his views were not only confirmed, but it appeared that he had underestimated his intergalactic distances by a factor of two.

Island universes: Each galaxy is a self-contained system of stars, distanced by millions of light-years from the other systems; our own galaxy is but one island universe among countless others.

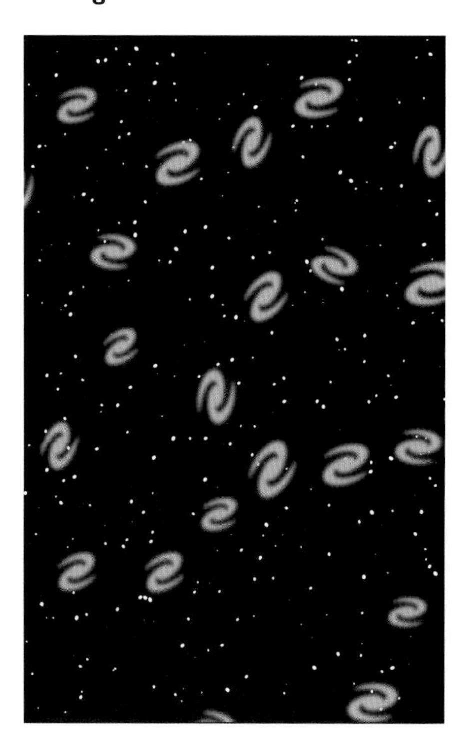

ISLAND UNIVERSES

The understanding of the galaxies as island universes permitted a new approach to the problem addressed by Hubble, that of mapping the universe in three dimensions. At first Hubble thought he had identified a large region in the sky where no galaxies appeared, but this was later explained as an optical aberration caused by dust in

Above: A galaxy cluster—Abell 2218—seen through the Hubble telescope. Abell 2218 lies two billion light-years from Earth. It is so massive that it bends the light of objects behind it. This is known as gravitational lensing and causes the arcs seen in this image.

Opposite: The Hubble telescope is named after the "second Copernicus."

deep space. It then appeared that the distribution of galaxies was uniform throughout space when taken on a large scale. This observation was later extended by the English astronomer E.A. Milne into a "cosmological principle" that the universe is uniform in all directions. Later work on galactic mapping since Hubble's death has in fact revealed significant clusters or major groups of galaxies, which has raised a question about this principle of uniformity. Whether these clusters have formed as galaxies have drawn together is still undecided.

A SECOND COPERNICUS

Hubble's most historic discovery came in 1929, and it related to the spectra of the galaxies. Their light was found to be shifted to the red end of the spectrum—the characteristic sign of a receding light source: Therefore these galaxies, which were already immensely distant, were actually moving away from Earth. Moreover, Hubble was able to show that there is a constant ratio between the distance and the velocity: The further away they were, the faster was their speed. Some of the galaxies observed by Hubble appeared to be moving at velocities of up to one-seventh of the speed of light. Nor was there any reason to suppose that this movement was only away from the Earth. The Earth could not possibly be imagined to be the center of the universe, and the cosmological principle suggested that the movement would be seen from any chosen point. The conclusion seemed inescapable that the galaxies were all rushing away from each other at enormous speed, and that the entire universe was therefore expanding.

This discovery was likened to a second Copernican revolution, for in place of an eternal, unchanging, motionless cosmos there now appeared a universe of intense, explosive movement. So strange was this discovery that even Hubble himself had some doubts about his findings. He wondered if some unexplained optical effects might be at work, so that the universe might be truly static. It is important to notice that everything in the universe is not expanding: The galaxies are not expanding internally, although they are in internal motion. The movement of the star Sirius that William Huggins found in 1868 (see Volume 7, page 55) was the motion of that star in relation to other stars and not part of the cosmic recession. The work of Hubble, building on the insights of Curtis and others, showed that it is the galaxies that are the large-scale units of the universe, and it is their distribution that is the key to the cosmic structure.

EDWIN POWELL HUBBLE (1889–1953)

- Astronomer and founder of modern cosmology.
- Born Marshfield, Missouri.
- Attended the University of Chicago to study mathematics and astronomy.
- 1910–13 Rhodes Scholar at Oxford studying law.
- Returned to the University of Chicago studying astronomy, then held a research post (1914–17) at its Yerkes Observatory.
- Did military service during World War I.
- 1919 Returned to astronomy to study nebulas at the Carnegie Institution's Mount Wilson Observatory.
- Discovered that spiral nebulas are independent stellar systems. Studied the velocity of galaxies.
- 1929 Published his discovery concerning the expansion of the universe: Galaxies recede at speeds that increase with distance. The "Hubble constant" is a measure of this rate.
- 1990 The Hubble Space Telescope launched on space shuttle *Discovery* was named after him.

Big Bang: The Exploding Universe

Opposite: The Big Bang theory was developed in the 1940s and 1950s and dominates modern cosmological thought. This conceptual artwork depicts the Big Bang, the huge explosion that cosmologists believe created the universe. It occurred about 12–15 billion years ago, although the exact figure is uncertain. The theory is based on the fact that the visible universe is still expanding outward from a central point and on the discovery of background microwave radiation thought to be an afterglow of the explosion.

Even before the extraordinary results of Hubble's observations became known, a number of physicists had been studying the effect of some of Einstein's ideas on cosmology. The Russian Aleksandr Friedmann (1888–1925) and the Belgian Georges Lemaître (1894–1966) argued that the universe must be nonstatic (in other words, constantly developing and changing) and that the curvature of space must be increasing with time. Hubble's work offered clear observational proof of this, and it raised two overwhelming questions: If the universe was really expanding, what had it expanded from, and what was it expanding into?

EXPANDING SPHERE

To take the second question first, the relativity of space to matter and light could now be understood in concrete terms: As the galaxies moved into ever-more remote empty space, so their presence defined that space and with it the structure of the universe. There could be no absolute space divorced from the material bodies that mark it out. The paradox that the universe was at the same time finite but without boundaries now made sense, and so did the curvature of space, for the universe was now imagined to be an expanding sphere in which the distances between all points was simultaneously increasing. The surface of the Earth or any other sphere has the same property: It is finite but without boundaries, and the galaxies could be visualized as points on the surface of an expanding balloon. But if we take this expansion back in time, if we imagine the film being rewound, how had it begun? If matter is everywhere receding, it seemed logical to suppose that it was once much closer together, and Lemaître introduced the quantum concept into the discussion. If we go back in time, fewer and fewer quanta (units) of energy will be found, until the whole universe is packed into a single quantum, an atom of unimaginable density. In this single point, Lemaître suggested, some instability had arisen causing an immense explosion that was the starting point of the expanding universe. This was the first appearance of the Big Bang theory of creation, although Lemaître was not able to give it any precise mathematical form or to suggest a cosmic time scale.

GEORGE GAMOW (1904–68

- Physicist.
- Born Odessa, Ukraine.
- Attended Leningrad University.
- Researched physics at various universities, including Göttingen, Copenhagen, Cambridge.
- 1931–34 Professor of physics at Leningrad University.
- Moved to the U.S. to become professor of physics at George Washington University (1934–55).
- 1948 With Ralph Alpher and Hans Bethe devised the "Big Bang" theory of the creation of the universe.
- Researched molecular biology, working on the order of nucleic acid bases in DNA chains. Made a major contribution to the understanding of DNA by discovering the "codes" within the proteins.
- Successfully wrote many books on popular science.
- 1956–68 Professor at Colorado University.

MODERN MYTH?

A much fuller and more detailed theory of the mechanics of the Big Bang came from the Russian-born George Gamow (1904-1968) in 1948, in a paper that has become famous in scientific history as the

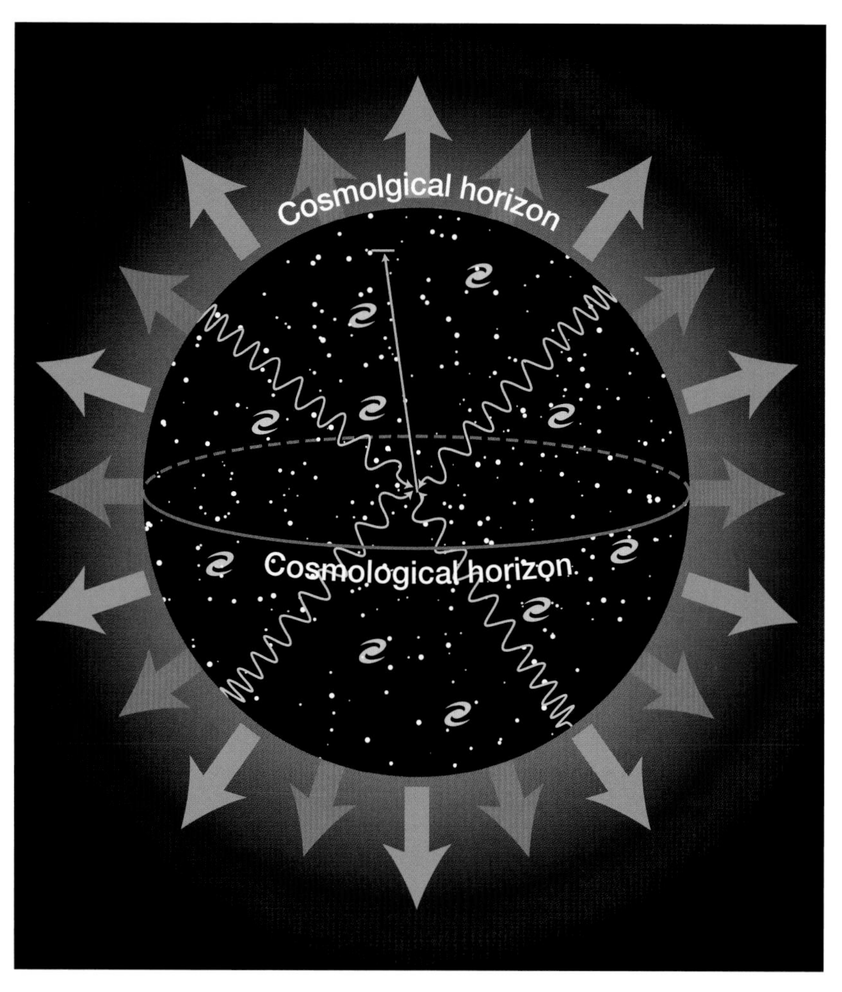

Above: Light coming from the horizon of the universe (red arrows) is red-shifted to infinity because those galaxies are receding; as the universe expands, the horizon expands even faster (blue arrows), bringing new sights into view.

Below: The aftermath of a supernova. The central ring is an expanding doughnut of material blown off by the explosion. The central spot is a new neutron star.

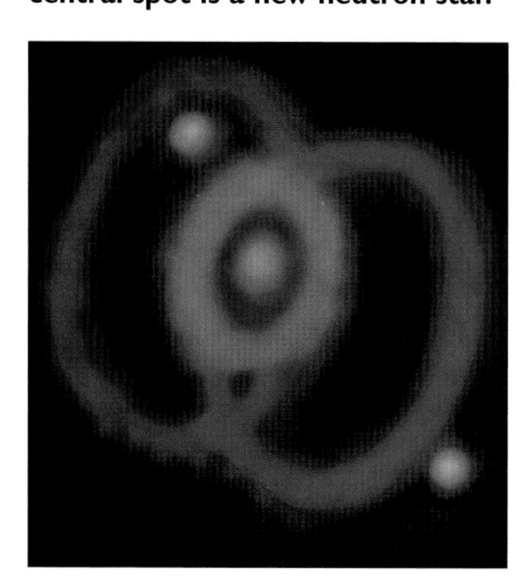

"Alpha-Beta-Gamma" paper because its coauthors were Ralph Alpher and Hans Bethe. Gamow himself later wrote a fuller account in his book *The Creation of the Universe* (1952). He suggested that all the protons, neutrons, and electrons in the universe had come into being at the moment of creation at unimaginably high temperatures, and that the structurally simple elements hydrogen and helium had been formed at a very early stage by nuclear aggregation. The heavier elements had, he believed, been formed later in the interior of stars, a view that later came to be universally accepted. Gamow tried to place the expansion of the universe within a time frame, calculating that some 17 billion years would be needed for the universe to reach its present size and shape. He made a prediction of the greatest importance, namely, that the radiation from the initial explosion should still be present permeating the universe even after such a vast passage of time.

The Big Bang theory dominates modern cosmological thought; indeed, it can be seen as the modern equivalent of old creation myths. Like the ancient myths, it is an explanation only in a limited sense: It offers a narrative of certain events, events that show the operation of some transcendent power. It satisfies the need for some explanation of the overwhelming mystery of how the universe has come to exist; but the ultimate how and why of these events are still beyond the reach of understanding. In the ancient myths it was a god who shaped Earth and stars out of chaos; but where did the god and the chaos come from? In modern thought it is the laws of physics that have shaped the universe of matter, but who formulated those laws? The only rational, nonreligious answer is that they are inherent in matter itself. If matter is to exist, it must exist in an ordered state. But why should matter exist at all? There is always another question that can be asked that neither science nor myth can answer. It is interesting to see that many astronomers and physicists have once again started to speak of God, although using the word in a quite impersonal way. They speak of God as the unfathomable power that arranged the universe from its birth, and that has determined what the laws of physics are. This impersonal God is proving useful as a summary of all that cannot be explained or rationalized any further, but just is.

The Elements of the Universe: B2FH

A further factor entered into astronomers' attempts to map the universe with the discovery of the interstellar medium—that the space between the stars and the galaxies is not empty but is filled with matter in the form of gases and very diffuse particles. In the late 1920s Robert Trumpler (1886–1956), a Swiss-born American astronomer, became puzzled by discrepancies in the size, brightness, and therefore the apparent distances of star clusters that he was studying. He conceived the idea that their light was being absorbed by an invisible medium, and that it was distorting our perception of them. In one particular area of the night sky, a strip about 20 degrees wide centered on the Milky Way, no nebulas at all were visible, so that it was at first believed by Hubble and others that no external galaxies existed in this "zone of avoidance." Trumpler understood correctly that the effects of this medium increased with distance, so that farther objects would appear fainter, leading to an overestimate of their distance. Those astronomers such as Harlow Shapley who had made estimates of the size of our galaxy accepted the validity of Trumpler's discovery and revised their figures downward by a large amount. Although the interstellar medium was very diffuse, it was clear that it must contain a huge volume of matter.

B2FH (GEOFFREY AND MARGARET BURBIDGE, WILLIAM FOWLER, FRED HOYLE)

Geoffrey Burbidge (1925–)
- Astrophysicist.
- Born Chipping Norton, England.
- Attended Bristol University, then University College, London.
- 1950s Collaborated with Fred Hoyle on the astrophysical consequences of antimatter.
- 1951 Moved to the U.S., first to Harvard, then Chicago, Mount Wilson, Palomar, and California.
- 1957 In collaboration with his wife Margaret, Hoyle, and William Fowler published a paper of theoretical research on nucleosynthesis—nuclear physics as applied to astrophysical circumstances,
- 1967 With Margaret published an important work on quasars,
- 1970 Highlighted the "missing matter" problem of galaxies.

continued on page 53

SPACE DUST

The dust in space was far from being a negative discovery, for it provided an answer to the problem of the birth of the stars. Hertzsprung and Russell had concluded that stars evolve over time. Next, Eddington and later Hans Bethe showed how this might happen on the atomic level. But where did the matter come from from which the stars evolved? Some stars were very old, and some were much younger, suggesting that there was a "life cycle" and a source of their material. Some years before the work of Hubble, before it had been decided that nebulas were star systems, the strange forms of the spiral-armed galaxies had attracted much attention: How could they possibly have arisen?

In 1919 the Cambridge physicist James Jeans (1877–1946) explained in theoretical terms that a spherical mass of gas would contract under the force of gravity, then start to spin until it became unstable. It would then throw out filaments of matter from its edges, which could then condense into spirals, exactly as seen in these nebulas. When huge volumes of gas were found to

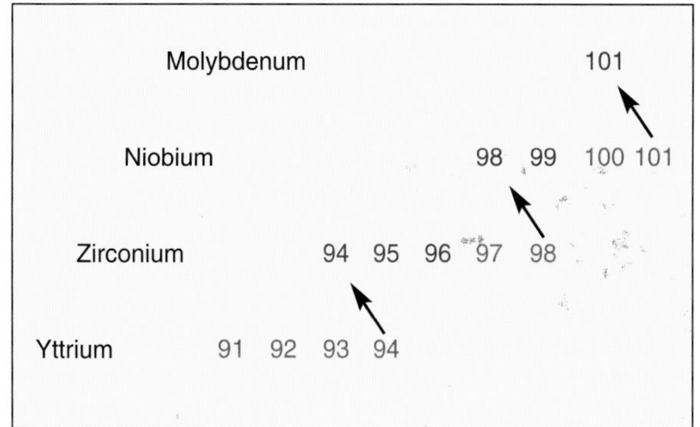

Heavy elements being created inside stars. Free neutrons are captured by atomic nuclei, which are then transformed into new elements. The red numbers are unstable isotopes, the blue are stable.

exist in interstellar space, added force was given to this theory, which in general terms is still believed to be correct. It was then discovered that where the interstellar medium was densest, it could be subject to spectral analysis, showing that it contained light gases, mainly hydrogen, but also nitrogen, oxygen, and dust particles: The interstellar space was found to be full of huge quantities of basic chemical elements. The gases were not evenly distributed, but were to be found in patches, some relatively dense. On the principle outlined by Jeans, it seemed that the birth of stars would occur when these gases condensed under the force of gravity and became sufficiently hot to start nuclear reactions.

EVOLUTIONARY THEORIES

From the 1930s onward theories concerning the evolution of stars were guided by a deepening understanding of atomic physics as observed here on Earth. That elements could transform themselves one into another by losing or capturing protons and neutrons was well known, but it applied to the lighter elements only. The process of star formation could not apparently account for the existence of all the heavier elements that are found on Earth, and that had been found in Sun's spectrum as long ago as the 1860s by Kirchhoff and Bunsen (see Volume 7, page 37).

This problem was studied by many leading astrophysicists, and in 1957 there appeared a paper entitled "The Synthesis of Elements in the Stars" written jointly by four scientists, Margaret Burbidge, Geoffrey Burbidge, William Fowler, and Fred Hoyle. This historic paper has therefore become known in the scientific world as B2FH, and it set out the physical rules for the creation of the elements past hydrogen and helium, which, its authors argued, was taking place continuously inside stars. The nuclear reactions inside the stars were constantly releasing free neutrons, which were built into heavier and heavier elements, a process that they plotted in precise mathematical order. This process continues as long as the star is active; but toward the end of its life a star will collapse when its material is exhausted, and it will explode in one of several ways.

The most spectacular and visible way is for it to become a nova or a supernova, a bright, violently exploding star, such as the one that Tycho watched in 1572 (see Volume 5, page 6). What is happening in a nova is that the material of a star, the chemical elements, is being ejected in atomic form back into the diffuse interstellar medium, where the cycle of star creation can begin once again. The B2FH paper was a landmark not only in astrophysics, but in our whole understanding of the universe, for it showed that a vast physical cycle is at work in the stars that has produced all the chemical elements that constitute our Earth and everything on it, including ourselves.

Opposite: A supernova—an exploding star that returns its chemical elements to interstellar space, to be recreated into new stars.

Margaret Burbidge, née Peachey (1923–)

- Astronomer specializing in galaxies and quasars.
- Born Davenport, England.
- Attended University College, London, and became interested in astronomical spectroscopy.
- 1948–51 Assistant director of London University observatory.
- 1951 Moved to the U.S. to Yerkes Observatory, then Caltech, then to the University of California.
- 1964–90 Professor of astronomy, University of California.
- 1972 Worked as director of the Royal Greenwich Observatory for a year
- 1979–88 Director of the Center for Astrophysics and Space Science at San Diego.

William Fowler (1911–95)

- Physicist and founder of the theory of nucleosynthesis.
- Born Pittsburgh, Pennsylvania.
- Studied at Ohio State University and California Institute of Technology.
- 1936 PhD from Caltech for work on radioactive nuclides.
- 1946 Appointed professor at Caltech
- Work concerned measuring nuclear reactions at low energies. Established the existence of the postulated excited helium state.
- After 1957 B2FH continued working on stellar nucleosynthesis and solar neutrino flux calculations.
- 1983 Shared the Nobel Prize for physics with Subrahmanyan Chandrasekhar.

Fred Hoyle (1915–)

- Born Bingley, England.
- Schooled at Bingley Grammar School and Emmanuel College, Cambridge.
- 1945–58 Taught mathematics at Cambridge.
- 1948 He and two colleagues advocated the "steady state" of the universe, that the universe is uniform in space and unchanging in time—a now discredited theory.
- Worked on investigating supernovas and the re-cycling of second generation stars from the exploded matter of earlier stars.
- Plumian Professor of Astronomy and Experimental Philosophy at Cambridge between 1958 and 1972.
- A prolific author of fiction and scientific works.
- 1972–78 Professor-at-large for Cornell University.

Problems and Conflicts in Cosmology

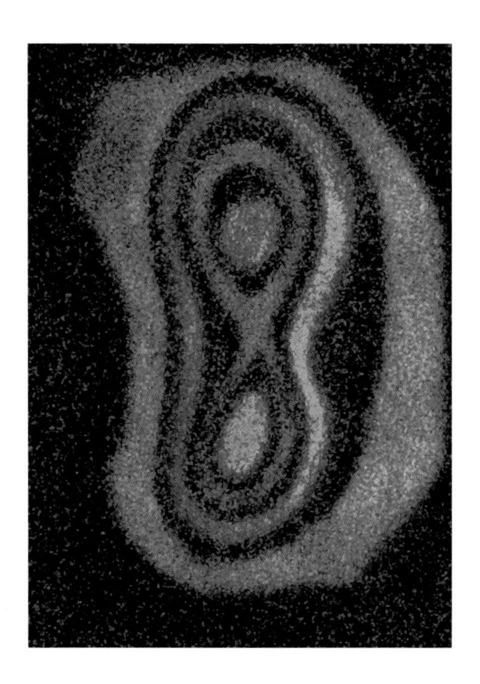

The quasar, above, appears to distort the light from a nearby galaxy, and its red-shift is identical to that of the galaxy, showing that it lies at a cosmological distance.

The idea that the stars are engaged in a vast cosmic process, recycling the chemical elements in a life cycle extending over billions of years, inevitably appears to raise questions about the Big Bang theory of the universe. The authors of "B2FH," and in particular Fred Hoyle, were never reconciled to the Big Bang model; indeed, it was Hoyle himself who coined the phrase Big Bang as a joke, although it was subsequently accepted by the whole scientific community. Instead, Hoyle, working with two Austrian physicists, Thomas Gold (1920–) and Hermann Bondi (1919–) , devised the "steady state" concept of the universe—that matter was being continually created in interstellar space. They were not able to explain precisely how this happened, but as they pointed out, neither could the Big Bang theory; and it was surely no more extravagant to claim that matter was being continuously created than that it had all come into being at one instant.

STEADY STATE

According to the steady state theory, the universe has no beginning and may have no end; it therefore removed some of the philosophical difficulties of the Big Bang, above all the problems of imagining what had existed before the moment of creation, and what had caused it. They argued that the Big Bang envisaged a time when the laws of physics did not apply, and then an arbitrary moment when they came, inexplicably, into existence, which they found absurd. There were several grave difficulties with the steady state model, however: If matter was being created, then the birth of galaxies, the building blocks of the universe, should be observable somewhere; but it was not. Nor could the dramatic recession movement of the galaxies be explained. But the event that undermined the steady state theory came from the new technique of radio astronomy.

RADIO WAVES

Radio astronomy had emerged by accident in the 1930s, when Karl Jansky (1905–50), an engineer for Bell Laboratories, was asked to investigate interference that was hampering transatlantic radio signals. Jansky built a large antenna and, turning it to the sky, found constant interference with a recognizable pattern over a period of 24 hours. Jansky realized that it was the Earth's rotation that was imposing this pattern, and that the radio waves must come from outside the Earth: It was the stars that were constantly bombarding the Earth with invisible radio signals, just as they emitted visible light.

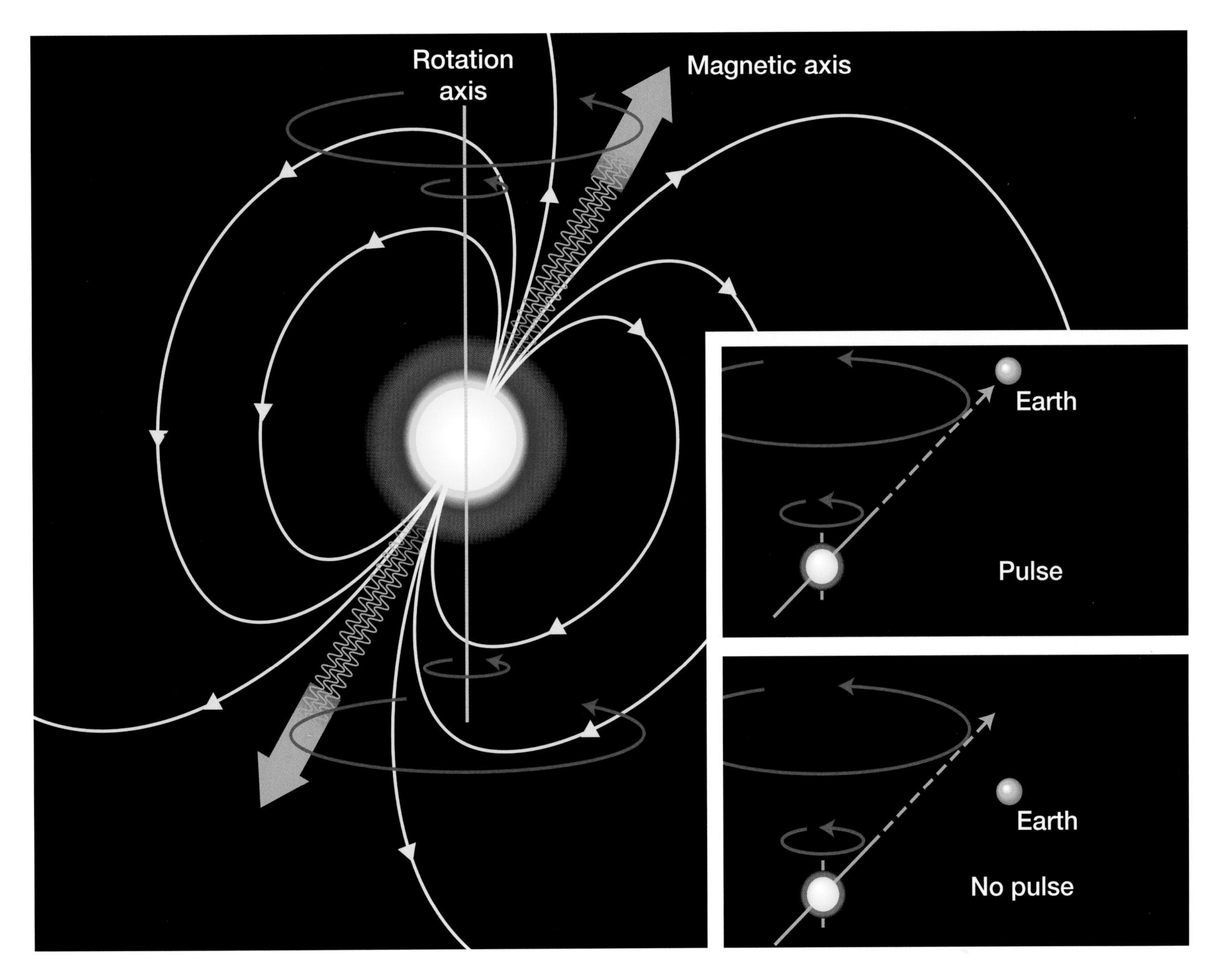

A pulsar's magnetic field spins in space around its axis of rotation. Radiation is beamed along the magnetic axis and will strike the Earth if suitably positioned.

After World War II this technique was developed to analyze stellar radiation and build up a map of celestial objects quite independent of optical telescopes. Radio telescopes were the means by which cosmic background radiation was discovered. George Gamow (see above, pages 48–49) had predicted that a radiation echo from the Big Bang should still be found permeating the universe, and in 1965 Arno Penzias (1933–) and Robert Wilson (1936–), two radio astronomers, found that every part of the sky, even a seemingly empty region, is giving off short-wave radiation at a constant low temperature of slightly under 3° Kelvin. This radiation was entirely even and uniform, not associated with individual objects, but permeating the whole sky. It was accepted as being the echo of the Big Bang, which Gamow had predicted, and the steady state theory had few supporters after this discovery.

STRANGE CHARACTERISTICS

Radio telescopes have been responsible for broadening our knowledge of the universe far beyond the visible stars. In 1967 the first pulsar was identified by Jocelyn Bell (1943–) at Cambridge in England. It was detected as a fast-flashing radio beacon of a kind so

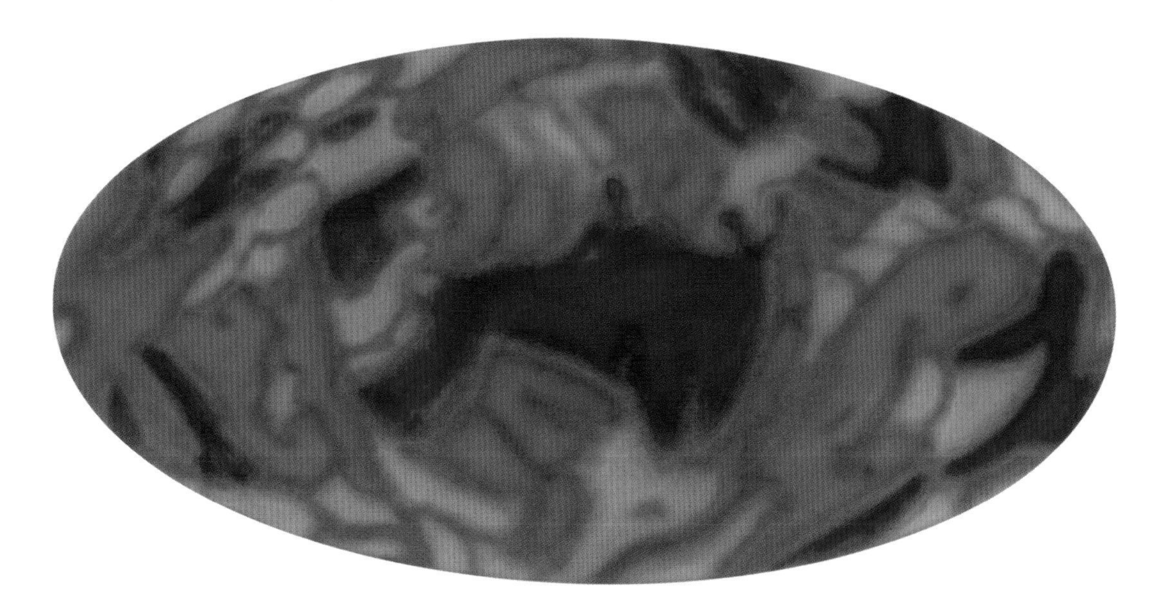

Cosmic background radiation, a crucial piece of evidence for the Big Bang theory; the red areas are hotter, the blue cooler, but the distribution is uniform.

novel that at first it was thought to be a communication from beings in deep space. Soon others were found, however, and it was concluded that pulsars were the remains of collapsed stars only a few miles in diameter, yet more massive than the Sun and spinning wildly, sometimes several times per second. Pulsars can be seen with optical telescopes, but their strange characteristics could never have been identified except through their radio signals.

Even stranger are the quasars, first detected through radio emissions, but later located optically. Quasars are bodies that are no more than a light-year in size, but have a tremendous luminosity equal to that of a whole galaxy with a diameter of 100,000 light-years. This permits them to be seen at cosmic distances of more than one billion light-years; in fact, the most striking thing about quasars is that they do not seem to exist at less than about three billion light-years distance, and most are much further away.

At the cosmic scale the viewing of distant objects is also a measure of time: When we see a quasar that is ten billion light-years away, we are seeing an object as it existed ten billion years ago. The clear implication is that quasars may not exist any longer, that they were an intermediate stage in cosmic evolution, and many astronomers believe that they are embryo galaxies.

In present-day astronomy the hunt is on for perhaps the strangest object known to astrophysics, the black hole. It has been suggested that in certain circumstances a star may collapse to such a dense state that no radiation, not even light, can escape from its surface. Matter approaching this dark region will be sucked into it by gravity and become invisible. For obvious reasons no black hole has ever been positively seen, but certain points of high radiation turbulence may be the edges of black holes. They are of great interest to physicists because they may represent areas of the universe where the laws of physics do not apply and may therefore hold the key to the nature of the universe before the Big Bang.

BIG SQUEEZE?

The age of the universe and the size of the universe are two interrelated questions. In the mid 1930s, after the meaning of Hubble's discovery had become clear, a size of two billion light-years for the universe was suggested by Englishman Arthur Eddington (1882–1944). But work on the evolution of stars and galaxies implied that their individual ages were greater than that. The advent of radio telescopes and the opening of the giant 200-inch optical telescope at Mount Palomar in 1949 both revealed galaxies at greater distances than two billion light-years, so once again the scale was extended. The current estimate is around 15 billion light-years in extent and 15 billion years for the age.

The two are inseparable, because when we see across 15 billion light-years of space, what we see is light or radiation that left its source 15 billion years ago. But there is no real reason to believe that this is a final figure, since cosmological thought is still very fluid. Astronomers cannot agree on the central question—whether the universe will continue to expand forever, or whether it will one day slow, halt, and go into reverse-collapse—the so-called big crunch or big squeeze. There is a tendency among today's astrophysicists to speak as though many of these fundamental problems were on the verge of being answered or had even been answered already. It should be remembered that ideas such as the Big Bang are still only theories, however widely they may be believed at present. It may be that further Copernican revolutions in our understanding of the universe are waiting to happen in the future.

Jodrell Bank's history is typical of many radio telescopes. It began in 1945 when Bernard Lovell needed a quiet site to observe cosmic rays. A quiet observing site was required, and Manchester University's botanical station at Jodrell Bank, 20 miles south of Manchester, was the ideal location. The Mark I Telescope—a giant at 250 feet—helped track down the quasars. Jodrell Bank was used in early space tracking activities. The second, Mark II, telescope was completed in 1964. In 1976 work began on the building of an 83-mile array of telescopes tied by microwave radio links to Jodrell Bank. In the early 1990s the array, now called Merlin, was extended to include a new 105-foot telescope at Cambridge and other improvements made to significantly improve its sensitivity.

The Birth of the Solar System

Opposite: Computer artwork of the planets solar system, seen from the Earth's surface. The planets are (clockwise from upper left) Saturn, Jupiter, Mars, a cloudless Venus, Uranus, and Mercury. The Earth is unique in possessing vast oceans of liquid water. Jupiter, Saturn, and Uranus are gas giants composed mainly of hydrogen and helium.

Below: False color image of the Sun showing the corona.

In 1906 an American astronomer and a geologist, F.R. Moulton (1872–1952) and Thomas Chamberlin (1843–1928), announced a startling new theory of the origin of the solar system. They suggested that at some time in the remote past, a second star had wandered close to the Sun, and that its gravitational pull had dragged a stream of matter out of the Sun, and the stream had then cooled to form the planets. This was a radically new idea, for until that time the nebular hypothesis of Laplace (see Volume 6, page 11) had dominated scientific thought. The Moulton-Chamberlin theory was widely believed, and one of its implications was that the solar system, having arisen from an extraordinary and perhaps unique event, might itself be unique.

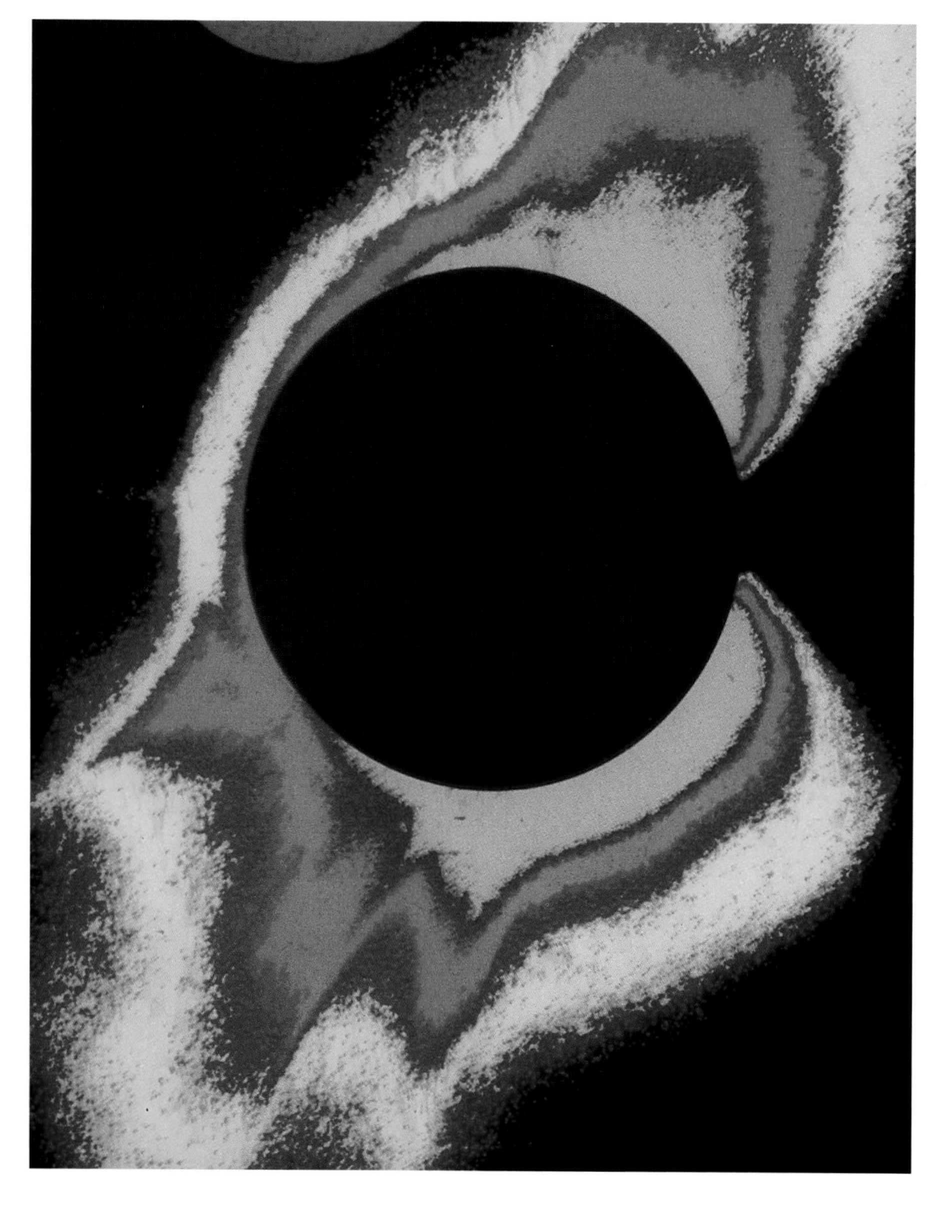

In the 1920s and 1930s eminent scientists such as James Jeans and Arthur Eddington considered that this might indeed be true, for no other planetary systems had been observed surrounding other stars. This is still true, and the history of the solar system is still a matter of theory only: Strangely, modern astronomers seem more certain about the origin of the universe as a whole than of our own planet and its immediate neighbors. In the nineteenth century geologists and naturalists such as Lyell and Darwin had argued that the age of the Earth must be measured in hundreds of millions of years. This conflicted with the best estimates of physicists like Lord Kelvin that the Sun itself could be no more than a few million years old, otherwise it would burn itself out.

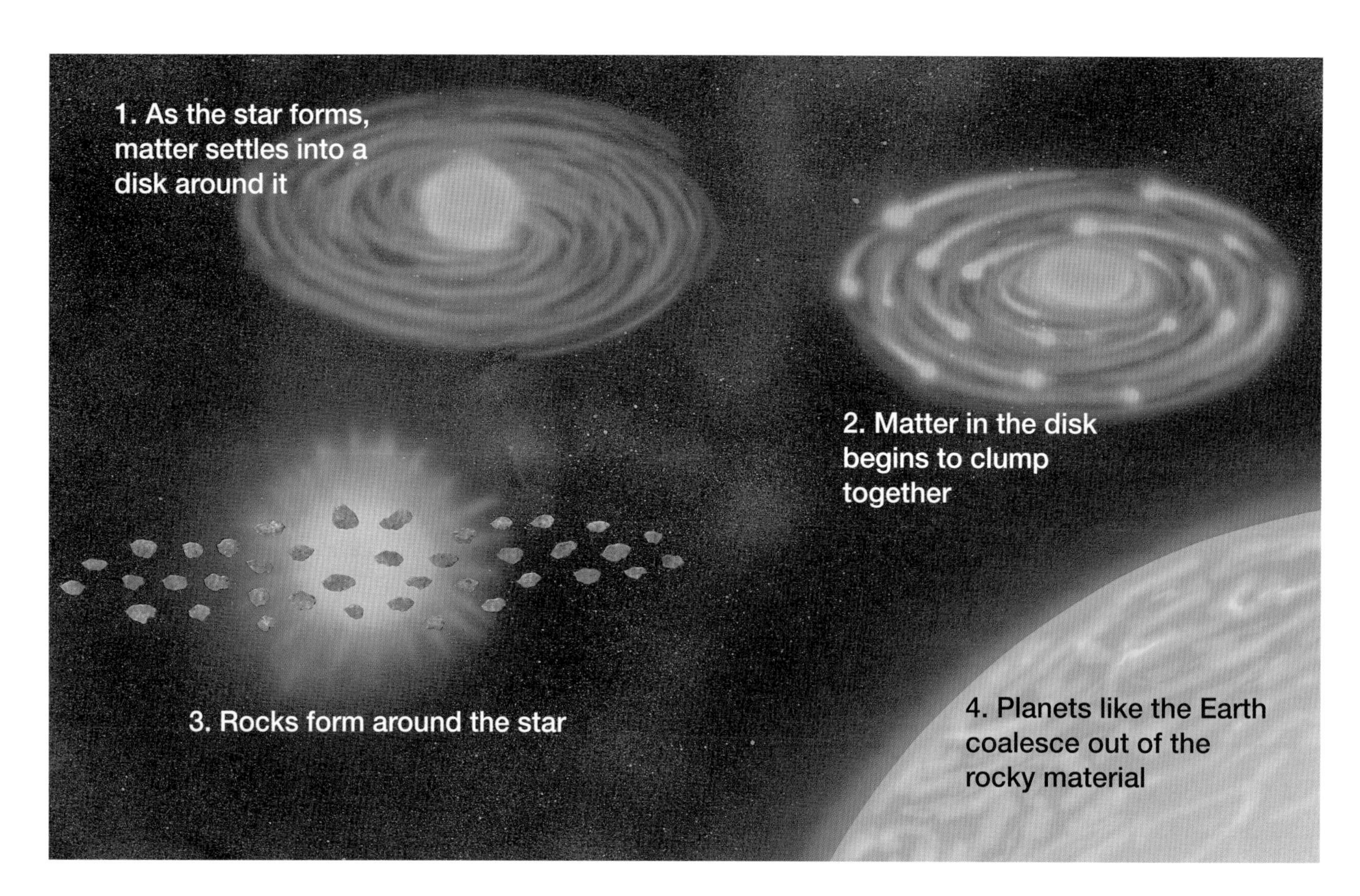

Above: The modern theory of planetary origins: Interstellar matter coalesces (bonds) under its own gravity, and the cooler, outer regions form planets.

Above Right: False color image of Callisto, one of the 16 satellites of Jupiter and one of the four so-called "Galileans"—the others are Io, Europa, and Gannymede. They are called this because they were discovered and observed by Galileo in January 1610. He called them the "Medicean planets" in honor of the great Italian Medici family.

METEORITE SAMPLES

The discovery of nuclear energy as the source of the Sun's power solved this problem, while techniques of dating through measuring radioactive decay enabled Earth scientists in the 1960s to estimate the age of the Earth as around 4.5 billion years. This figure was confirmed when geologists were able to analyze meteorite samples and then lunar samples. Astrophysicists also agreed that the Sun is approximately halfway through its life cycle of 10 billion years: It seemed therefore that the entire solar system had originated at the same time. The Moulton-Chamberlin theory was generally rejected, and scientists turned back to the earlier nebular model. Since we know from spectroscopy that all the heavier elements that make up the Earth—iron, magnesium, carbon, and so on—are also present in the Sun, it seemed that Laplace had been right after all: Gases had condensed to form the Sun and the planets at the same time.

SOLID PARTICLES

The nebular hypothesis was revived by the German Carl von Weizsäcker (1912–) in 1945. Beginning with the protosun surrounded by a disk of rotating gas, Weizsäcker argued this mass would break up into smaller vortices or eddies, which in turn condensed to form the planets. The differing composition of the planets, from the small, solid, rocky, iron-cored inner planets, to the massive but very cold gaseous outer planets, was determined by temperature differences as the vortices were spread further from the Sun.

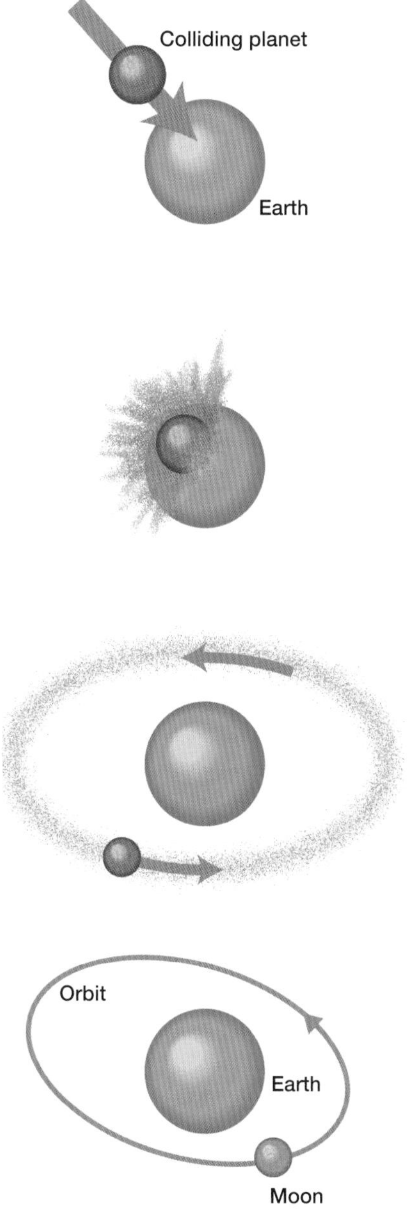

Above: The origin of the Moon is still unknown: Was it formed at the same time as the Earth, or was it torn from the Earth by a cosmic collision?

Subsequent theories gave a greater role to the solid particles that are found in the interstellar medium, ice-covered particles of silicon, carbon, and other elements. It is now believed that they came together under the force of the protosun's gravity, forming "planetesimals," which grew in size until they formed the planets. The asteroids and other bodies in the solar system—including some of the satellites of Jupiter and perhaps even the planet Pluto—may be large surviving planetesimals. The dynamics and the chemistry of this process are still the subject of theoretical investigation.

From time to time astronomers have suggested new catastrophic theories as playing a part in the history of the solar system, for example, that the Moon was split from the Earth by a collision with another large body. Unlikely as this sounds, it would explain some geological similarities between the two bodies. In complete contrast, it has also been argued that the Moon was originally a body wandering from a distant part of the solar system, which was "captured" by the Earth's gravity.

The origin of the moons of Jupiter, Saturn, and Uranus are even more mysterious, some rocky, some icy, some volcanic, and many of them so different from their parent planets. If the origin of the solar system were more certain, we would be better able to say how likely it is that other such systems exist. It is a paradox of modern astronomy that despite penetrating the depths of space, our knowledge of the birth of our own solar system, and its possible uniqueness, is still so uncertain.

The Earth's Magnetism: James van Allen

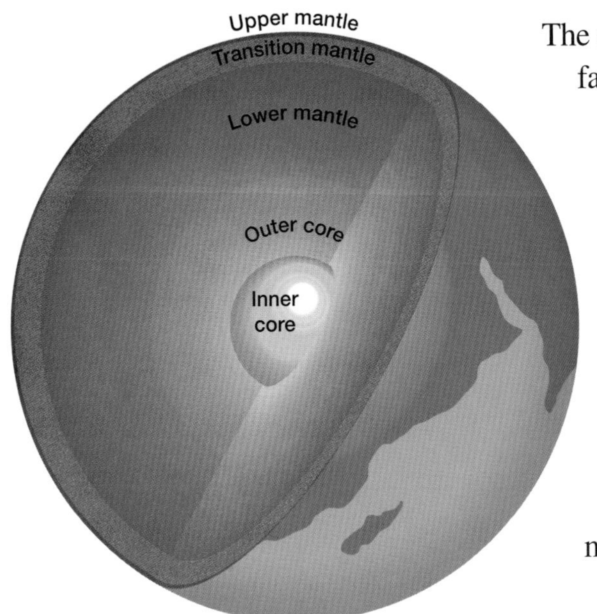

The existence of the magnetic force is one of the oldest scientific facts known to humans, and it has been one of the most practical, for the invention of the magnetic compass in the late Middle Ages revolutionized the art of navigation. But what was the source of this magnetic effect? One of the earliest theories was that the compass was pointing toward the pole of the sky, until the Elizabethan scientist William Gilbert announced his discovery that the Earth itself was a vast magnet with twin poles (see Volume 5, pages 33–34). This idea was universally accepted, but for hundreds of years no scientific explanation of the Earth's magnetism was forthcoming until more precise methods of geophysical measurement were developed in the early twentieth century.

Above and Below: The Earth's magnetic field arises from its fluid internal structure (above), mapped by seismologists in the early twentieth century.

SHOCK WAVES

Between 1895 and 1910 the English seismologist Richard Oldham (1858–1936) made careful studies of earthquakes in India, where he was the head of the Geological Survey. By comparing data from different monitoring stations, Oldham found that the seismic waves appeared to travel at varying speeds to places opposite the focal point of the quake. From this he deduced that the Earth's interior was not uniform, but that it contained a central core that was denser and more elastic, and through which the waves therefore traveled faster. In later years the study of the speeds and paths of these shock waves through the Earth has made it possible to map the Earth's interior, showing its various solid and liquid layers. The existence of these layers has also provided the explanation for the Earth's magnetism, namely, that the Earth acts as a geomagnetic dynamo. Fluid motions in the liquid outer core, which consists largely of molten iron at high pressure and high temperature, set up an electric current, with its associated magnetic field; these motions are continually sustained by the Earth's rotation.

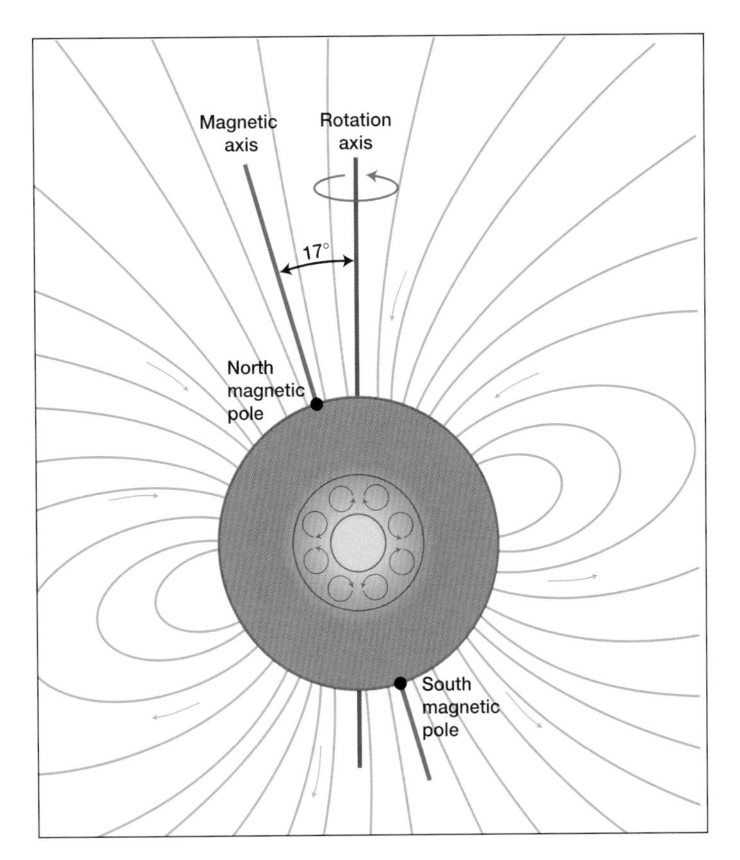

VAN ALLEN BELTS

The Earth's rotation and therefore its magnetic field have far-reaching effects on climate and therefore on life; indeed, in the region around this planet magnetism creates a zone known as the magnetosphere in which the magnetic field extends some 40,000 miles into space. In this zone the Earth's magnetic force deflects most of the charged particles of radiation that come from the Sun. In the direction of the Earth's travel the magnetosphere resembles the bow wave of a ship moving through water, while behind it the tail stretches out much further beyond the Moon's orbit. The most significant regions within the magnetosphere are the two van Allen belts, identified by the American physicist James van Allen in 1958.

Van Allen (born 1914) was a naval scientist who transferred to rocket research after the end of World War II. A series of experiments carried out from the earliest orbiting satellites showed that a considerable amount of solar radiation leaked into the magnetosphere and became concentrated in two large, paired, crescent-shaped zones, one inner and one outer, that curved down toward the Earth's magnetic poles. Their function was confirmed when several small nuclear explosions were set off 300 miles above the Earth in order to release very energetic particles into the magnetosphere. Monitoring of this radiation showed that it had indeed been captured in the van Allen belts.

The Earth's magnetic fields, working through the van Allen belts, clearly protects the Earth from harmful radiation from space. We cannot say that this is its purpose, but it provides a striking example of balance and symmetry in nature: If this one aspect of the physical world were slightly different, life on Earth would be impossible.

JAMES VAN ALLEN (1914–)

- Pioneer space physicist.
- Born Mount Pleasant, Iowa.
- 1935 Graduated from Wesleyan College in Iowa.
- Worked at the Department of Terrestrial Magnetism at the Carnegie Institution of Washington, studying photodisintegration.
- 1942 Moved to the Applied Physics Laboratory at Johns Hopkins University. Developed a rugged vacuum tube and helped develop radio proximity fuses for weapons, especially antiaircraft projectiles used by the U.S. Navy.
- 1942 Commissioned into U.S. Navy and sent to Pacific to test the fuses.
- 1946 As director of high-altitude research at John Hopkins investigated the Earth's upper atmosphere.
- 1951–85 Professor of physics and head of the physics department at the State University of Iowa.
- 1958 Participated in the development of Explorer I, the U.S.'s first satellite. It also carried his cosmic ray detector.
- Satellite observations showed two rings of radiation—magnetically charged particles speeding around the Earth trapped by its magnetism—they are known as the van Allen belts.
- Involved in the first four Explorer probes, the first Pioneers, Mariners, and the orbiting geophysical observatory.

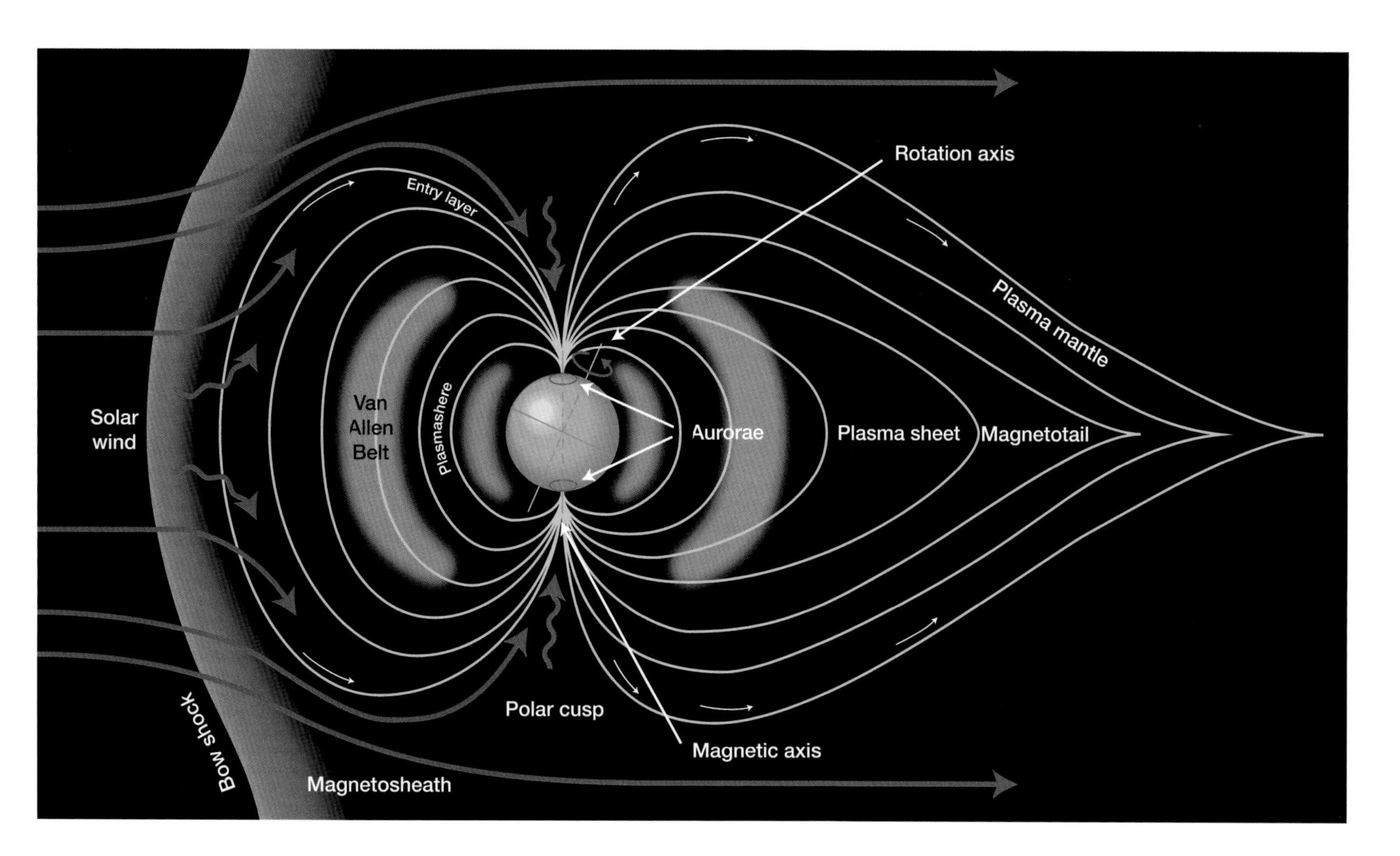

The van Allen belts poised like a shield in space protecting the Earth from solar radiation.

Astronomy:
What Remains to Be Discovered?

The uncertainty about the origin of the solar system is a sobering reminder of the puzzles and mysteries that still await answers in the realm of astronomy. There has been a century of extraordinary discovery about the stars and the cosmos, but every discovery seems to lead to the unexpected or even to the inexplicable. Claims that we are on the verge of final answers about the origin and workings of the universe need to be treated with a great deal of skepticism: If astronomers cannot explain where our own planet and its immediate neighbors came from, can we really believe their accounts of the birth of the entire universe? There is a cluster of fundamental questions in cosmology that are currently impossible to answer, and that prevent us even now from forming a reliable picture of our universe.

OLDER UNIVERSE?

First, there is Hubble's constant, the relation between the velocities of the remote galaxies and their distance from us. Hubble's constant should express the rate at which the universe is expanding; but what is Hubble's constant? Hubble himself estimated that it was 93 miles per second per million light-years, from which Hubble deduced that the universe was some 2 billion years old. However, modern cosmologist have revised this drastically downward, to between 9 and 18 miles per second per million light-years. This slower rate means that the universe must be much older, possibly 10 billion to 20 billion years; but this is a huge range, and current observations cannot make it any more precise.

People's understanding of the universe has been revolutionized several times; the Big Bang theory may not be the last word, and its long-term fate is totally unknown.

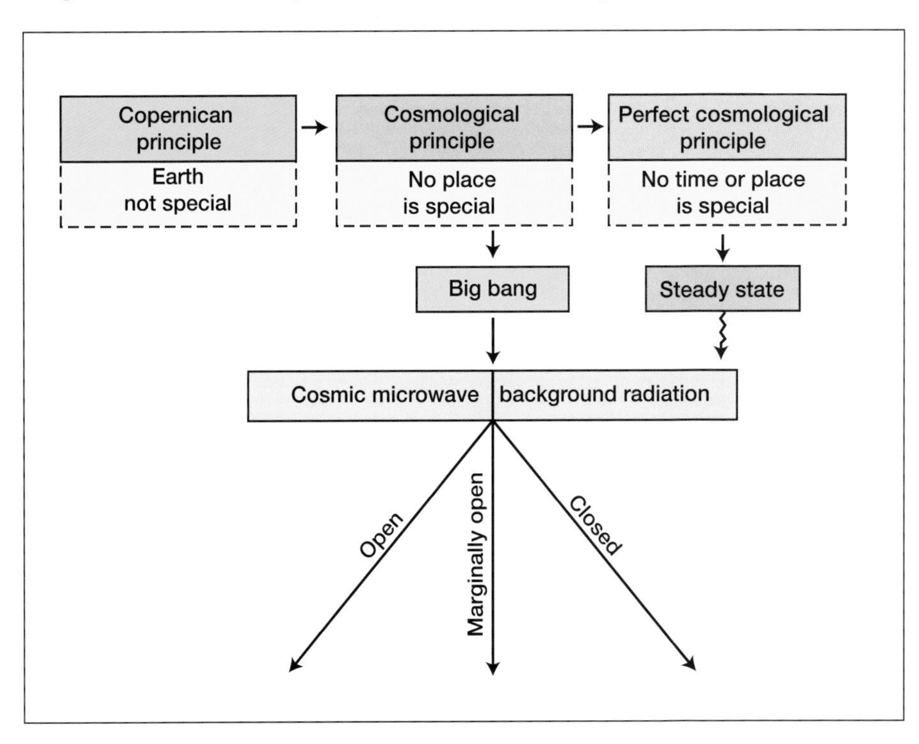

ORIGIN OF GALAXIES

Then there is the question of the galaxies, the basic building blocks of the universe: Do stars come together into galaxies under the influence of their gravity, or are galaxies formed as communities of stars out of vast clouds of interstellar dust? The former seems unlikely, for no stars have ever been observed that are not part of a galaxy; yet the process of the origin of galaxies is not observable either. Meanwhile, the distribution of the galaxies is still unmapped. Within a certain range of vision a number of major clusters have emerged, but no symmetrical pattern: Does the universe have a large-scale structure, or are the galaxies scattered randomly through space?

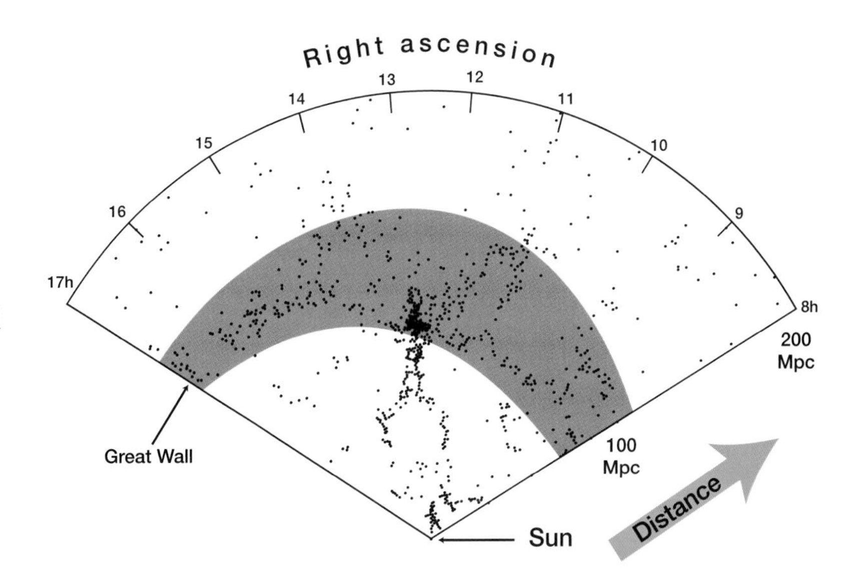

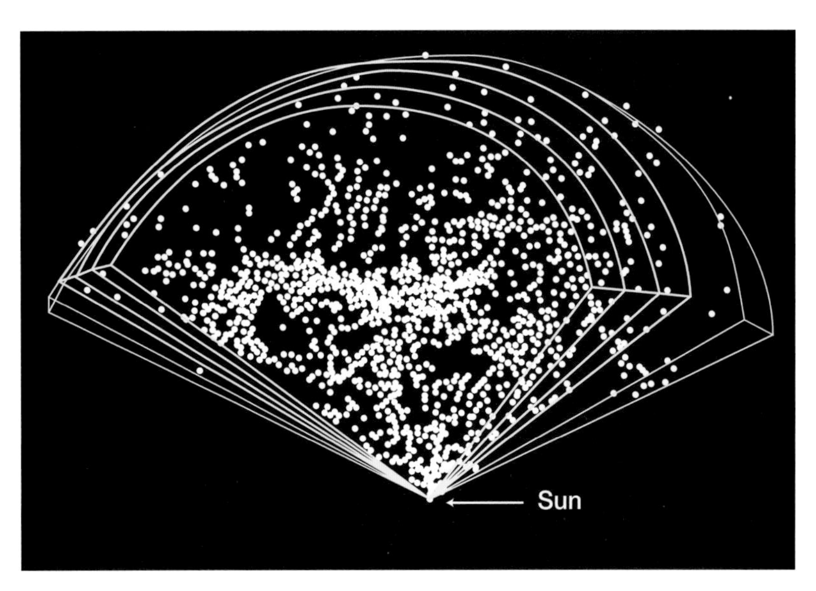

Is there a large-scale structure to the universe? Plots of galaxy positions reveal strange concentrations, but no clear pattern.

EXPANDING UNIVERSE

The most intriguing problem is whether the universe will continue to expand or not. Will the explosive force of the Big Bang one day exhaust itself and be defeated by the gravity between the galaxies, so that its expansion will slow and then halt, and all the matter in the universe will collapse back into itself?

Cosmologists claim that, in theory, they should be able to answer that question by comparing the mass of the universe with the velocity at which galaxies are moving apart. This calculation has proved impossible, however, mainly because of the "dark matter" that is now believed to fill space.

As long ago as the 1930s the Swiss astronomer Fritz Zwicky (1898–1974) tried to calculate the mass of a galaxy from its luminosity, using the Sun as a standard measurement. Zwicky was astonished to find that galaxies appeared to possess 50 times more mass than their luminosity would require, and this result was confirmed by other observers too. The only conclusion to be drawn was that galaxies were filled with matter so diffuse as to be invisible. This major unknown makes it impossible to say if the universe is "open" or "closed"—whether it will go on expanding forever or will one day contract into the dense state that existed before the Big Bang. This may seem a totally academic question, since in either case the end of the universe lies billions of years in the future, yet it is a burning issue in science because of humankind's consuming need to map the universe it lives in. But it is clear that the unknown in the universe still enormously outweighs the known. It will be time to put forward grand "theories of everything" when we know what everything is.

Conclusion: Power and Responsibility in Science

Science and its sister technology are indisputably the most powerful forces shaping the world and the way we live. For this reason it is our duty to think about the role of science in society and to question where it is leading us. There is no aspect of life into which mechanical and electrical machines do not enter. The environment in which humankind once lived was, as with all other living species, the natural environment. But the tools, machines, and systems that we have created mean that this is no longer true: The environment in which people now live is that of technology.

A CIVILIZATION OF MACHINES

Technology has insulated many people from natural forces such as cold, hunger, darkness, distance, and physical labor. It has been said that technology is the art of so arranging the world that we never have to experience it directly; in other words, technology now forms an environment that we can manipulate at will. But if technology distances us from nature, does it not also distance us from human nature, too, from ourselves? The classical objection to a civilization based on machines is that people becomes the slave of the machine instead of being its master, and this is exactly what has happened to millions of people in modern industrial society.

Above: Mars taken through the Hubble space telescope in 1997.

Right: The clouds of Venus from Pioneer.

NO LONGER PURE SCIENCE

Until perhaps a century ago it might have been possible to draw a distinction between pure science and practical technology. Machines were built by engineers or technicians as they sought better means of transportation or communication or production, and these technicians were rarely scientists in any academic sense. But now, in the age of electronic and atomic power, technology is led by pure science, and the machines that we use in everyday life function according to principles that have been discovered by the most advanced physicists.

"Natural philosophy"—the search for nature's fundamental structures and forces—is not the province of the isolated scholar, but now has immediate consequences in physics, chemistry, medicine, or genetics.

We saw in the case of the Manhattan Project that modern science can easily acquire a political and military dimension because new discoveries in pure science can open up the possibility of new weapons of immense destructive power. The possession of nuclear weapons by two great rival powers dominated world politics for almost half a century following World War II, but it could be argued that it was the very threat of those weapons that prevented war between them. In addition to the nuclear technology itself the search for techniques to guide and direct these weapons was a major driving force in electronic technology.

COMMERCIAL MOTIVE

The application of electrical and then electronic technology to domestic and personal use revealed another great dimension to science—the commercial motive. New machines, or improved versions of old machines, offered wealth to the inventor and the manufacturer. The modern industrial and consumer worlds are always in search of novelty and improvement, and they do not arise by accident but as a result of scientific research.

The point about the involvement of science in politics and in commerce is that they have both strengthened enormously the momentum of science, to a point where its progress is impossible to stop and perhaps even to control. It is, it seems, impossible to foresee the consequences of scientific and technical change. Knowledge is relentlessly pursued for its own sake, new powers are developed, and new machines are built because they are possible. The consequences only emerge much

Above: Voyager I's view of Jupiter and the great storm swirling around the lower hemisphere.

later. And yet it if there is one great law of human history, it is the law of unpredictability: Events and forces that are set in motion produce results decades or centuries later that were never foreseen at the outset. Managing change has been identified as the greatest challenge facing modern people, and this may be true; but it hides the problem of where change comes from: Who initiates change and why? The greatest single sources of change are science and technology, so it is vital that we ask where these changes are leading. Do we have choices about these changes, or does science now have a momentum that overrides the individual?

Above: Color-enhanced view of Saturn taken in 1980.

Right: Full Moon taken in 1972.

FASTER AND BETTER?

A familiar example is the rise of computers. When the use of computers began to spread rapidly through the worlds of commerce and government in the early1980s, people said that their influence would always be limited because "we could always switch them off." In fact, there is now virtually no aspect of social and personal life that is not regulated by computers, and to switch them all off would be unthinkable. One-third of the working population of the Western world now work by operating computers all day every day.

This may be good or it may be bad, but the question is who decided that this should happen? Was it the scientists who set out the theory of binary logic in the 1950s, or those who pioneered miniature electronic circuitry in the 1960s, or those who saw the huge commercial potential in the 1970s? Who now decides how much of our lives is handled through electronic databases? Will a future generation arise that interacts with the outside world exclusively through electronic screens?

Science is largely responsible for the culture of improvement that now dominates our lives, the idea that everything must steadily become better, faster, more efficient. We become dazzled by new techniques and lose sight of the value of what we are doing. Henry David Thoreau, a nineteenth-century critic of progress, said that our inventions are apt to be mere toys, improved means to unimproved ends.

TAKING CONTROL

Science began as an intellectual quest to understand the laws of nature. But the unveiling of those laws has placed in our hands powers that have led in unforeseen directions. The medieval legend of Doctor Faustus told of a man to whom knowledge was more important than life itself: He sold his soul to the devil in exchange for limitless knowledge of nature's secrets. It was given to him, but in the end he had to pay the price, and the devil collected his due. Although told in the language of medieval parable, the Faust story is the perfect symbol of humankind's ceaseless desire for knowledge and its destructive consequences.

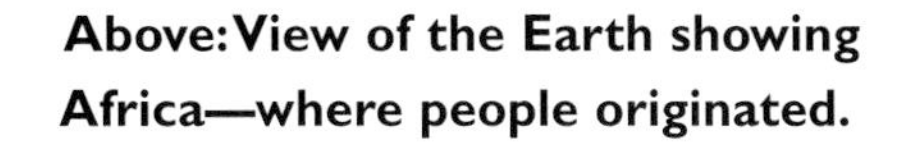

Above: View of the Earth showing Africa—where people originated.

Almost three centuries later, at the height of the first industrial revolution, the English poet and artist William Blake likewise prophesied that science would prove to be "the tree of death." But surely it should be within our power to restore to science its intellectual role, but to avert or control its potentially destructive powers. These questions will be examined again in the context of ecology and the life sciences in Volume 10.

Glossary

Artist's impression of a binary star system consisting of a black hole (lower right) and a red giant star (upper right). A stream of gas is being pulled from the atmosphere of the red giant by the immense gravitational field of its companion. As the gas spirals down into the black hole, it forms a flat accretion disk that heats up and emits X-rays (bright white spot at center). For this reason this type of system is known as an X-ray binary. Twin jets of material are being ejected from the poles of the black hole. A second, similar system is seen in the background at left.

astrophysics a branch of astronomy dealing with the behavior of celestial bodies.

atomic physics branch of physics dealing with the atom.

black hole hypothetical celestial body with gravity so strong that no light can escape.

cosmic background radiation a relic of the Big Bang, this radiation—also known as the microwave background—is a weak radio signal that is detectable in every direction like cosmic white noise.

ecology branch of science dealing with the relationship of organisms and their environment.

electron an elementary particle with a negative electrical charge.

fission the splitting of the nucleus of an atom, thus releasing large amounts of energy.

fusion the union of atomic nuclei to form heavier nuclei and resulting in the release of enormous amounts of energy.

geophysics branch of science dealing with the physical processes occurring in the Earth.

Kelvin temperature scale for thermodynamics on which absolute zero is 0° Kelvin; in temperature intervals 1° Kelvin is the same as 1°C.

kinetic relating to the movement of bodies and energy associated with it.

nebula a cloud of dust or gas in space.

parallax the angular difference in direction of a celestial body measured from two points on the Earth's orbit.
photons a quantum of electromagnetic radiation.
plate tectonics geological theory that the Earth is divided into a small number of plates each floating over the mantle; many of the Earth's volcanic eruptions and earthquakes occur where they meet.
pulsar the remains of a collapsed star only a few miles in diameter, yet more massive than the Sun and spinning wildly, sometimes several times per second.

quantum any of the very small parcels into which units of energy are subdivided.
quark subatomic particle thought to come in pairs.
quasar celestial body, no more than one light-year in size, but with luminosity equal to a whole galaxy with a diameter of 100,000 light-years.

radioactivity the quality of emitting energetic particles by the disintegration of atomic nuclei.
radio astronomy astronomy dealing with radio waves received from space.
relativity theory that as speed increases, time moves more slowly, until finally, at the speed of light, time stops altogether.

seismologist scientist who deals with earthquakes and other vibrations of the earth.
spectral analysis analysis of the spectrum by a spectroscope.
spectrum arrangement of components of a complex color in order of energy—producing, when put through a prism, the visible rainbow colors (red, orange, yellow, green, blue, indigo, violet) and a range of other colors invisible to the naked eye.
spectroscope instrument used in spectroscopy, involving a slit to produce a parallel beam, a prism to disperse the different wavelengths, and another tube to observe the results. See photograph in Volume 7, page 53.

Captions for page 72.
Top: Albert Einstein.

Below: Astronaut Dave Scott photographed next to the Lunar Roving Vehicle by Jim Irwin during the Apollo 15 mission. Apollo 15 landed in the Hadley-Appenine region on July 30, 1971.

Captions for page 73.
Top: Certificate awarded to the Curies.

Below: Colored satellite map of atmospheric ozone in the southern hemisphere on September 6, 2000. The ozone hole (green) over Antarctica (dark green) is at its 2000 maximum of 11 million square miles.

Below: Computer illustration of a pulsar in its nebulous supernova remnant. Pulsars are rapidly rotating neutron stars that cast out narrow beams of energy as they rotate. Any pulsar whose beam chances to cross Earth will appear to be flashing like a lighthouse. Pulsars rotate extremely fast, with periods ranging from hundredths of seconds to a few seconds. The pulse is visible from radio to X-ray wavelengths.

PERIOD 1900 1910 1920 1930 1940

WORLD EVENTS

1903 First powered flight by the Wright brothers.
1912 Sinking of the Titanic.
1914 Panama Canal opened.
1914–18 World War I.
1917 United States enters war.
1917 Russian Revolution.
1919 Treaty of Versailles redraws map of central Europe.
1919 Influenza kills millions worldwide.
1919 First transatlantic flight.
1919–33 Prohibition era.
1924 Stalin comes to power in Russia.
1933 Hitler comes to power in Germany.

SCIENCE

1895 Röntgen discovers X-rays.
1896–1905 Curies research radiation.
1897 Thomson discovers the electron.
1900 Planck's quantum theory.
1905 Einstein publishes theory of relativity.
1905–15 Cushing maps out endocrine system.
1910 Rutherford's theory of atomic structure.
1913 Bohr's quantum atomic theory.
1913 Hertzsprung-Russell diagram.
1915 Wegener's theory of continental drift.
1915 Morgan's theory of genetic mutation.
1919 Rutherford splits the atom.
1920 Shapley-Curtis debate on scale of the universe.
1924 Hubble shows nebulas are separate galaxies.
1925–30 Fisher and Haldane develop population genetics.
1925 Dart discovers *Australopithicus* fossils.
1926 Eddington shows that stars are powered by nuclear energy.
1926 Schrödinger's wave atomic theory.
1927 Heisenberg's uncertainty principle.
1928 Fleming discovers penicillin.
1930 Hubble announces model of expanding universe.
1938–39 Nuclear chain reaction theory.

ART & CULTURAL EVENTS

1890 onward Post-Impressionist art: Gaugin, Matisse, Picasso.
1905 Functional architecture: Wright, Bauhaus, Le Corbusier.
1912 Abstract painting begins with Kandinsky.
1913 Stravinsky's "Rite of Spring."
1915 Experimental novelists and poets: Proust, Kafka, Lawrence, Eliot, Joyce.
1920s Jazz age.
1920s Modern American novel: Hemingway, Steinbeck, Faulkner.
1920s Hopper's paintings.
1920s Manhattan skyscrapers.
1930s Color films, TV broadcasts.

1940 1950 1960 1970 1980 1990 2000

1939–45 World War II.
1941 Japanese attack on Pearl Harbor.
1941 Germany declares war on United States.
1945 Atomic bombs dropped on Nagasaki and Hiroshima.
1947 India gains independence from Great Britain.
1948 State of Israel declared.
1950–53 Korean War.
1953 Death of Stalin.
1955 onward Decolonization of Africa.
1961 First man in space.
1961 Berlin Wall put up.
1962 Cuban missile crisis.
1964–72 Vietnam War.
1964 Civil rights movement in the United States.
1986 Chernobyl disaster.
1989–90 Collapse of communism in Europe.
1994 Majority rule in South Africa.

1940–50 Dobzhansky & May's "New Synthesis" in genetics.
1948 Gamow: Big Bang theory.
1950–52 Hershey & Chase show that DNA determines biological characteristics.
1950–75 Fieldwork by the Leakeys show humans evolved in East Africa.
1953 Miller-Urey experiments on the origin of life.
1953 Crick and Watson discover DNA structure.
1954 Salk's polio vaccine.
1957 B2FH—mechanism of element building in stars.
1958 Discovery of the Van Allen belts.
1962 *Silent Spring*—beginnings of an ecology movement.
1963–65 Wilson's theory of plate tectonics.
1963 Quasars discovered.
1964 Gell-Man proposes quark theory of subatomic structure.
1965 Penzias and Wilson discover cosmic background radiation.
1967 Pulsars discovered.
1967 First human heart transplant.
1970 Microprocessors inaugurate computer revolution.
1975 Hole in ozone layer discovered.
1980 Alvarez proposes cretaceous extinction theory.
1988 Global warming identified.
1995 Human genome project in progress.

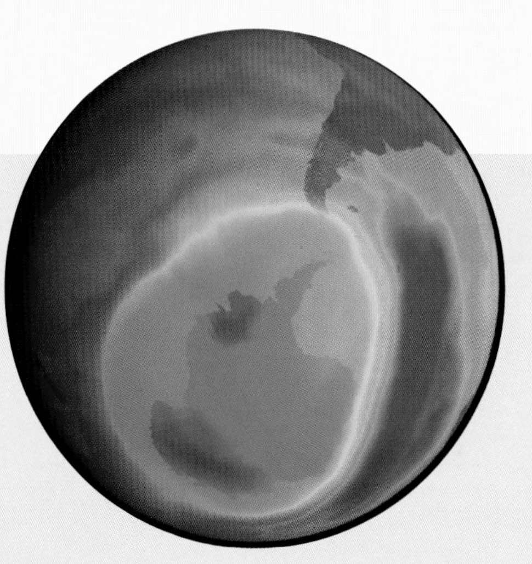

1940 onward Hollywood dream factory.
1950 Electronic music introduced.
1950s American drama: Williams, Mailer, Albee.
1956 Birth of rock and roll.
1960 New cinema in Europe: Bergman, Fellini, Godard.
1960s Youth counterculture.
1970s Feminism.

Resources

FURTHER READING

There is a wealth of books published on the history of science, particularly biographies of great scientists. The following list includes many large works that contain many further resources.

Adams, F.D.: *The Birth and Development of the Geological Sciences*; Dover Publications, 1955.
Bowler, P.: *Evolution: The History of an Idea*; University of California Press, 1998.
Bowler, P.: *The Norton History of Environmental Sciences*; W.W. Norton & Co., 1993.
Boyer, C. & Merzbach, U.: *A History of Mathematics*; John Wiley & Sons Inc., 1989
Brock, W.H.: *The Fontana History of Chemistry*; Fontana Press, 1992.
Butterfield, H.: *The Origins of Modern Science*; Free Press, 1997.
Clagett, M.: *Greek Science in Antiquity*; Dover Publications, 2002.
Cohen, I.B.: *Album of Science: From Leonardo to Lavoisier*; Charles Scribner's Sons, 1980.
Cohen, I.B.: *The Birth of a New Physics*; W.W. Norton, 1985.
Crombie, A.C.: *Augustine to Galileo: The History of Science AD400–1650*; Dover Publications, 1996.
Crombie, A.C.: *Science, Art and Nature in Medieval and Modern Thought*; Hambledon, 1996.
Crosland, M.: *Historical Studies in the Language of Chemistry*; Heinemann Educational, 1962.
Eves, H.: *An Introduction to the History of Mathematics*; Thomson Learning, 1990.
Gillispie, C.C. (ed.): *Concise Dictionary of Scientific Biography*; Charles Scribner's Sons, 2000.
Gillispie, C.C.: *Genesis and Geology*; Harvard University Press, 1996.
Hallam, A.: *Great Geological Controversies*; Oxford University Press, 1983.
Ihde, A.J.: *The Development of Modern Chemistry*; Dover Publications, 1983.
Jaffé, B.: *Crucibles: The Story of Chemistry from Alchemy to Nuclear Fission*; Dover Publications, 1977.
Jungnickel, C. & McCormmach, R.: *Intellectual Mastery of Nature: Theoretical Physics from Ohm to Einstein*; University of Chicago Press, 1986.
Koyré, A.: *From the Closed World to the Infinite Universe*; The Johns Hopkins University Press, 1994.
Kuhn, T.: *The Copernican Revolution: Planetary Astronomy in the Development of Western Thought*; Harvard University Press, 1957.
Lindberg, D.C.: *The Beginnings of Western Science*; University of Chicago Press, 1992.
Porter, R. (Ed.): *The Cambridge Illustrated History of Medicine*; Cambridge University Press, 1996.
McKenzie, A.E.E.: *The Major Achievements of Science*; Iowa State Press, 1988.
Morton, A.G.: *A History of Botanical Science*; Academic Press, 1981.
Nasr, S.H.: *Islamic Science—An Illustrated Study*; London, 1976.
North, J.D.: *The Fontana History of Astronomy and Cosmology*; Fontana Press, 1992.
Olby, R. (et al.): *A Companion to the History of Modern Science*; Routledge, 1996.
Parry, M. (ed.): *Chambers Biographical Dictionary*; Chambers Harrap, 1997.
Porter, R.: *The Greatest Benefit to Mankind: a Medicinal History of Humanity from Antiquity to the Present*; HarperCollins, 1997.
Roberts, G.: *The Mirror of Alchemy*; British Library Publishing, 1995.
Ronan, C.A.: *The Cambridge Illustrated History of the World's Science*; Cambridge University Press, 1983.
Ronan, C.A.: *The Shorter Science and Civilisation in China*; Cambridge University Press, 1980.
Selin, H. (ed.): *Encyclopedia of the History of Science, Technology and Medicine in Non-Western Cultures*; Kluwer Academic Publishers, 1997.
Uglow, J.: *The Lunar Men*; Faber and Faber, 2002.
Van Helden, A.: *Measuring the Universe: Cosmic Dimensions from Aristarchus to Halley*; University of Chicago Press, 1985.
Walker, C.B.F. (ed.): *Astronomy before the Telescope*; British Museum Publications., 1997.
Whitfield, P.: *Landmarks in Western Science: From Prehistory to the Atomic Age*; The British Library, London, 1999.
Whitney, C.: *The Discovery of Our Galaxy*; Iowa State University Press, 1988.

THE INTERNET

Websites relating to the history of science break down into four types:

- Museum sites that offer some history and artifact photography. This is often the easiest way to visit international sites or those of states too far away to get to in person.
- College or other educational establishment sites that often provide online learning or study resources.
- General educational sites set up by enthusiasts (often teachers) and historians.
- Societies or clubs.

Examples of these three types of website include:

Museums
http://www.mhs.ox.ac.uk/
Museum of the History of Science, Oxford, England.
Housed in the world's oldest surviving purpose-built museum building, the Old Ashmolean.

http://www.mos.org/
Museum of Science, Boston.

http://www.msichicago.org/
Museum of Science and Industry, Chicago.

http://www.lanl.gov/museum
Bradbury Science Museum, a component of Los Alamos National Laboratory.

http://www.si.edu/history_and_culture/history_of_science_and_technology/
Smithsonian Institution site.

http://www.sciencemuseum.org.uk/
National Museum of Science and Industry, London, England.

http://galileo.imss.firenze.it/
Institute and Museum of the History of Science, Florence, Italy.

http://www.jsf.or.jp/index_e.html
Science Museum, Tokyo.

Colleges or institutions
http://sln.fi.edu/tfi/welcome.html
Franklin Institute with online learning resources and study units.

http://www.fas.harvard.edu/~hsdept/
Department of the History of Science of Harvard University.

http://www.hopkinsmedicine.org/graduateprograms/history_of_science/
Department of the History of Science, Medicine and Technology at Hopkins.

http://www.lib.lsu.edu/sci/chem/internet/history.html
Louisiana State University provides excellent history of science internet resources and links.

http://dibinst.mit.edu/
The Dibner Institute is an international center for advanced research in the history of science and technology and located on the campus of MIT.

http://www.mpiwg-berlin.mpg.de/ENGLHOME.HTM
Max Planck Institute for the History of Science

http://www.princeton.edu/~hos/
History of Science @ Princeton.

http://www.astro.uni-bonn.de/~pbrosche/hist_sci/hs_sciences.html
History of sciences from Bonn University, Germany, including indexes on the history of astronomy, chemistry, computing, geosciences, mathematics, physics, technology.

Educational sites
http://echo.gmu.edu/center/
ECHO—Exploring and Collecting History Online—provides a centralized guide for those looking for websites on the history of science and technology.

http://www.wsulibs.wsu.edu/hist-of-science/bib.html
Provides reference sources in the form of bibliographies and indexes.

http://dmoz.org/Society/History/By_Topic/Science/Engineering_and_Technology/
Open Directory Project providing bibliography and links.

http://orb.rhodes.edu/
ORB—the Online Reference Book—provides textbook sources for medieval studies on the web. It includes the Medieval Technology Pages—providing information on technological innovation and related subjects in western Europe—and Medieval Science Pages, a comprehensive page of links to medieval science and technology websites.

http://www.fordham.edu/halsall/science/sciencesbook.html
This page provides access to three major online resources, the Internet Ancient History, Medieval, and Modern History Sourcebooks.

http://www2.lib.udel.edu/subj/hsci/internet.htm
The University of Delaware Library provides an excellent guide to Internet resources.

Societies
www.hssonline.org
History of Science Society provides for its members the History of Science, Technology, and Medicine Database—an international bibliography for the history of science, technology, and medicine.

http://www.chstm.man.ac.uk/bshs/
British Society for the History of Science.

Set Index